中国古代计量史

THE HISTORY OF ANCIENT CHINESE MEASURES AND WEIGHTS

丘光明 著
张延明 译

时代出版传媒股份有限公司
安徽科学技术出版社

This book is the result of a co-publication agreement between Anhui Science and Technology Publishing House (China) and Time AACP (Australia)

The History of Ancient Chinese Measures and Weights

By Qiu Guangming

ISBN: 978921816031

Published by:

TIME AACP
Expanding Horizons
www.aacpasia.com

An imprint of the ATF Ltd.
PO Box 504
Hindmarsh, SA 5007
ABN 90 116 359 963
www.aacpasia.com
Expanding Horizons

序

计量科学源于远古，发展于现代。中华文明绵延5000年，计量技术一直与社会生产、商品交换互相促进，并成为中国科学技术发展的重要基础。

中国古代计量技术的最大特点，是受到儒家礼乐制度的深刻影响。古代经典著作《尚书》云：“协时月正日，同律度量衡。”《礼记》云：“周公朝诸侯于明堂，制礼作乐颁度量而天下大服。”2000多年来，度量衡的发展与乐律有着千丝万缕的联系。秦始皇统一度量衡，以乐律为基础建立起来的单位制，经过《汉书》整理成文，代代相袭，至清代仍被康熙、乾隆所遵从，甚至延续至中华民国时期。同时，由于受古制的束缚，度量衡的改制与创新皆难以被接受，如宋朝学者李照奉诏制乐秤，提出将秦汉以来的16两1斤制改为10两1斤制，以1升水之量定1斤之重，然而，这一更加科学的制度未被采纳。尽管如此，中国的度量衡制度随着时间的推移持续地发展着，最终形成了一个博大精深的科学体系，从而在世界科学技术史上占有重要地位。这部中英文对照的《中国古代计量史》，便是从这个历史宝库中精选最具代表性的器物图片，并加以扼要论述，它从一个特殊的层面反映了中华民族的智慧和对世界文明的贡献，也是与各国计量史同行进行交流的极好图书。

本书编著者丘光明女士从事计量史研究，三十年如一日，不知疲倦地搜集历史资料，研究实物器件，发表了数十篇学术论文，出版了多部专著。丘女士之所以取得丰硕的研究成果，除了她的勤奋刻苦之外，也有一点家学渊源。她的祖父是清末代进士，官至保庆知府，也是知名学者和教育家；她的父亲是一位学富五车的文史专家。这一家庭背景为她从事学术研究并取得丰硕成果创造了非常好的条件。我衷心地期望她继续努力，为这个重要的事业做出更多贡献。

是为序。

Metrology originated from the ancient times and developed in the modern times. Throughout five thousand years of the Chinese civilization, metrological technologies have been advanced with social production and commercial exchanges, and have laid the foundation for the development of the Chinese sciences and technologies.

The most striking characteristic of the ancient Chinese metrological technologies lies in the fact that they were influenced profoundly by Confucian rites. It is recorded in the classics of *Shang Shu* that "setting the seasons in the calendar in line with days and months; and associating the measures and weights with the rites". In *Li Ji*, it reads, "in handling with the affairs of state, Zhou Gong had meetings with the feudal princes in order to regulate rites and measures and weights, thus winning the subjects". It is evident that for more than two thousand years, the development of the measuring and weighing systems has been associated with that of rites and temperament. Qin Shi Huang, the First Emperor, unified the measuring and weighing systems, and the units were set on the basis of the rites and temperament. Sorted out and put in order in *Han Shu*, these were carried out in the following dynasties, even observed by Kang Xi and Qian Long, emperors of the Qing Dynasty, and continued till the Republic of China. Meanwhile, being bound up in the ancient systems, efforts on innovation and reformation on the measuring and weighing systems met with difficulties and were unable to be adopted. For instance, Li Zhao, a scholar in the Song Dynasty, when receiving an imperial edict to produce steelyards, advocated to change the scale of 16 liang as 1 jin prevailing in the Qin and Han dynasties to that of 10 liang as 1 jin; and set the capacity of 1 sheng of water as 1 jin. Although this was more scientific, the then government did not accept it. The Chinese measuring and weighing systems, however, developed continuously and became broad and profound scientific systems. They also play an important role in the history of science and technology in the world. This book in Chinese and English consists of precise descriptions with pictures of the most representative objects from the historic treasuries. It is from this special facet that the collections in the book reflect the wisdoms of the Chinese people and their contributions to the world civilization. The book is also a masterpiece with which people may exchange ideas with those in other countries who share the same interest in the history of metrology.

Qiu Guangming, the author of this book, is a researcher on the measuring and weighing systems for more than 30 years. She has been collecting historical data and examining actual objects, which resulted in numerous research papers and books. There are domestic origins in her scholastic endeavors: Her grandfather was a famous scholar (Jin Shi), government official and educator in the Qing Dynasty, and her father was a renowned expert in literature and history. All these backgrounds contributed to her establishments and successes in pertinent researches. I expect wholeheartedly that she will continue to make endeavors and to contribute greatly to this important cause.

Wang Daheng (Endorsement)

Foreword

THE HISTORY OF ANCIENT CHINESE MEASURES AND WEIGHTS

前 言

中国古代以度量衡和时间为主要内容的计量技术，有着悠久的历史。早在父系氏族社会，度量衡和计时已是农业文明的基础。传说在黄帝时代已发明了以干支记日、月。继而尧命羲、和二人参照日、月、星、辰定历法。舜到东方巡视，在部落联盟议事时，商讨把四时之气节、月之大小、日之甲乙，度量衡之齐同，乐律声音之高低都统一起来。禹治理水患，划分九州，“身为度，称以出”，以人体建立度量衡标准。上述虽然都属后人追溯，却真实地反映了先民们的自然哲学观念。

计量制度的建立、单位标准的确定虽然都是人为的，但必须具有权威性。公元前221年，秦始皇用武力征服了各诸侯国，颁发了统一度量衡诏书，同时初步建立了一套完整的度量衡制度。后经汉代的改进、完善，成文于典籍而被历代遵循，奉为圭臬。此后每经改朝换代，都要探究古制之本，以确定当朝度量衡和计时单位标准。历代流传下来的器物不断被发现，其传承关系明确便是有力的证明。直至清代，无论是度量衡还是计时制度都是秦汉古制的沿袭，今犹陈列在北京故宫博物院太和殿前的鎏金铜嘉量和日晷便是有力的物证。

中华民族的祖先，通过长期的生产实践和天文观测，创造了里亩、尺寸、升斗、斤两等度量衡单位制和年、月、日、时、刻等计时单位。以当时先进的科学方法，制定了单位量值标准，不断完善测量器具和测量方法，使中华民族创造的物质财富和科技文明，都能在时空坐标上被定格记录下来，其量化的数据真实、可比。

中华悠久的文明史流传下来大量珍贵文物，其中有许多与计量有关的器物和文字资料，记录和讲述了一个个生动而有价值的故事。如考古学家曾统计过，在100多座春秋战国时期楚国的墓葬中，出土了数量不等的天平、砝码，它们是用来称量可切割的黄金货币的，反映了楚地盛产黄金、经济繁荣、商贾活跃的社会面貌。掌握着大量财富的王公贵族和豪商巨贾们，向往着死后升入天堂，继续过着荣华富贵的生活，天平、砝码便成为随葬品埋入地下。又如留存至今的“商鞅铜方升”，器壁刻铭详尽，其中“十八年”即器的制造年代在商鞅辅助秦孝公变法的公元前344年，是为了统一秦国度量衡而由官居卿相的商鞅亲自督造的。“十六尊（寸）五分尊（寸）壹为升”，说明当时已普遍使用“以度审容”。“齐遣卿大夫众来聘”，记录了当时重大的政治事件，同时也可能有两国共同商定有关统一度量衡的内容。公元前221年，秦始皇下诏书统一全国度量衡，又将诏书加刻在器的底部。一件量器所刻铭文，向后人讲述了秦国几百年的历史，它的重要意义远远超过了器物本身。秦始皇统一度量衡几乎是世人有口皆碑的历史功绩，秦权、秦量出土地域之广、数量之多，令人惊叹。据粗略统计，出土地域囊括了被统一的每一个诸侯国旧地，数量多达百余件。这些都展示了秦始皇统一度量衡的决心和雄才大略。

中国古代计量技术，在历代史籍中都有辑录。研究者根据文献记载，对照所能见到的器物，考释其铭文，测量其实际数值，模拟、复现其计量功能，使尘封的古老科技重现光彩。西汉末年，刘歆总结了前人的智慧，提出了一套系统的以黄钟、累黍定度量衡标准的理论，

后载入《汉书·律历志》。即以音频定长度,用累黍为介质加以复现。声与量皆为无形之物,量尚可通过度量衡器测量出来,而在古代,声是无法测量并保存的。如何让以黄钟定度量衡这一科学的设想得以实现,刘歆等先哲们通过反复试验后,提出以黍为介质,即选用中等大小的自然物“黍”,横排90粒得黄钟律管之长(9寸),加一寸即一尺。律管之积约容1200粒黍,与一龠的容量相当,二龠为一合(16.2立方分)。积1200黍之重约12铢,24铢即一两。这样便形成了一组参量公式,厘定出度、量、衡三个单位量。从此这一量化了的公式便成为其后历代制定当朝制度时可遵循的古代标准,既可以用度量衡来定律,又可以用律来校度量衡。近现代许多学者对此做过大量的理论推算和实物验查,证明这一公式是符合物理学原理的,在一定误差范围内也是可以复现的。

中国古代许多杰出的科学家在科学实践中因离不开“时空量”的测量,都关注和研究度量衡及时间计量。如战国时的著作《墨经》,首先讨论了等臂杠杆和不等臂杠杆各种平衡的状况,对衡器的制造作了理论上的分析。不足之处是尚未以应用数学或数学语言来表达这些平衡关系。三国时期的数学家刘徽注释《九章算术》时以“新莽嘉量”与三国时期·魏尺度、容量作比较,得出魏尺、魏斛分别增大了4.7%和2.6%的结论。南朝数学家祖冲之,经反复运算、研究后,得到了精确到小数点后七位数的圆周率(3.1415926~3.1415927)。并用它校核了“新莽嘉量”的设计数据后,指出其设计者——刘歆的数术“粗疏”。唐代在经过长期的实际运用后,把延续了1600多年的衡制中的“两、铢、累、黍”(非十进制),改进为“十钱为一两”。后又经过宋代进一步完善,推出了“两、钱、分、厘、毫”皆十进制的质量小单位系列。宋代的另一特点是,众多的天文、音律、算术家以及达官显宦都参与详考古今(宋)计量器具。如开国之初便有刘承珪精校朝廷收支银两的各种衡器,创制了精巧的小型杆秤——戥子,重新制颁了成套标准砝码。李照研制了“乐秤”(又称水秤),提出用1升水定1斤重的标准。司马光与范镇有关是“律生度”还是“度生律”的讨论,历时三十年,往返书信数万言。北宋科学家沈括著《浮漏仪》,将漏刻的结构及消除误差的各种措施一一记录下来。《宋史·舆服志》中对指南车、记里鼓车的内部结构、齿轮传动系统和机械装置作了详细的叙述。

中国近代对计量史的研究始于20世纪初,罗振玉、王国维、刘复、马衡、容庚、商承祚、唐兰、朱德熙等学术巨擘都先后发表了重要著作,对商鞅方升、秦权、秦量、新莽嘉量、历代尺度等作了大量考证和研究。1937年吴承洛著的《中国度量衡史》,是中国第一部度量衡通史专著。从20世纪50年代开始,随着考古事业的发展,古代计量器具陆续出土,历史、文物、考古、古文字专家先后发表了许多有分量的研究论文。20世纪70年代末,在国家计量总局及其直属单位的领导下,有了一支专业的研究计量史的队伍,此后有关计量史的论文、图书陆续出版,中国计量史成为科技史研究的一个分支,也逐渐受到国际上一定的关注。但由于汉语难学、难懂,曾有许多外国朋友来信、来访时提出,希望有一本英文版的中国计量史,简要地介绍一些有关资料供他们参考。于是笔者便产生了以图文并茂的形式编写一本《中国古代计量史》的想法。经过多年的努力,并得到国家质量技术监督检验检疫总局计量司的大力支持,终于写成该书。该书初稿完成后,又得到北京传世文化发展中心总裁李胜兵先生和安徽科学技术出版社的青睐,才被提到出版日程上来。

希望通过这本专著的出版,广大读者能给予中国计量史更多的关注,也希望引起更多国内外朋友对计量史研究的兴趣,更希望能与各国学者展开交流、切磋和合作,并盼望将来有一部世界性的比较计量史书籍面世。

丘光明

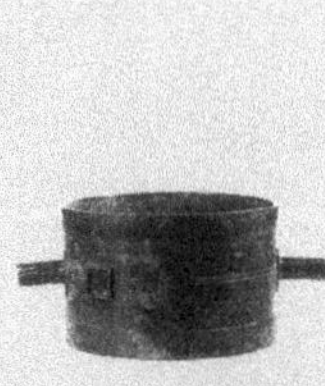

Foreword

In ancient China, there was a long-standing history in metrological technologies which encompassed time,measures and weights. As early as in the patriarchal society, time reckoning, together with measuring and weighing systems, built up the foundation of the agricultural civilization. According to legends, in the times of Huang Di (the ancestor of the Chinese people who had Yao, Shun and Yu as his successors), a technology was invented to record days and months by the Heavenly Stems and Earthly Branches. Yao ordered Xi and He to establish a calendar in accordance with the positions of the sun, the moon, stars and constellations. While taking inspection tour to the east, Shun presided over meetings with the union of tribes to discuss the unification of the four seasons, days in the months, measures and weights, and pitches in the temperament. While regulating rivers and floods, Yu divided the land into nine administrative regions (Jiu Zhou), and stipulated the lengths of the parts of his body as the standards in measuring and weighing objects. Although these are from popular legends, they reflect the natural philosophical concepts held by our ancestors.

In ancient times, although personalized, the establishment of the metrological systems and the formulation of the standards of units had to be authoritative. In the year of 221 B.C., Qin Shi Huang, the First Emperor, conquered the other states and issued an imperial edict on the unification of the measuring and weighing systems. A complete set of such systems was then initially established. After further innovations and improvements in the Han Dynasty, these systems were confirmed in written forms and then observed as the standards in the following dynasties. Whenever a change of dynasties took place, the original systems were re-examined to make sure that the standards of units were the same as before. An examination of the unearthed apparatus made in the successive dynasties offers clear evidence of such successions. In the Qing Dynasty, the measuring and weighing systems and time reckoning systems were set in accordance with those ancient systems prevailing in the Qin and Han dynasties. The Gilded Bronze Jia Liang and Sundial exhibited in the Forbidden City Museum serve as substantial evidence.

Throughout practices in production and astro-observations, ancestors of the Chinese people created such measuring and weighing units as li, mu, chi, cun, sheng, dou, jin, and liang; and time reckoning units as year (nian), month (yue), day (ri), hour (shi) and quarter (ke). In setting up the standards of units, advanced scientific methods were implemented, and the production and methods of measuring and weighing apparatus were continuously improved. These represent the material treasures and scientific and technological civilization created by the Chinese nation. They have been recorded throughout the history. The data are quantized and comparable.

In the long history of the Chinese civilization, valuable cultural and historic relics have been handed down in large quantities. Among the materials and written data, there are numerous records that tell us vivid and interesting stories. For example, archaeologists have had statistics that lots of balances and weights were unearthed from more than one hundred tombs of the Warring States. They had been used in the State of Chu for weighing gold and currencies. This shows that the State of Chu was abound in gold; the economy was prosperous; and commercial activities were lively. The unearthed materials demonstrated that the rulers and the rich who controlled the wealth had balances and weights buried with them, intending to maintain wealth and status in the netherworld. Another example is the Shang Yang Tong Fang Sheng, a bronze apparatus of capacity produced by Shang Yang, which is kept till present with carved descriptions in details, showing that it was made in the "Eighteenth Year (of Qin)" when Shang Yang was assisting Qin Xiao Gong the Prince in his Reformation in 344 B.C. This demonstrated clearly that in order to unify the measuring and weighing systems, it had to be done with the supervision of an official of high rank—as high as

Qing Xiang (equivalent to the rank of Prime Minister). The statement of "16 cun and 5 fen as 1 sheng" illustrated that the method of "determining the capacity by measures (of lengths)" was adopted. The statement of "Officials (Da Fu) from the State of Qi arrived in large groups" showed that it not only recorded a significant political event, but also indicated possible meetings at a high level to discuss and decide on unifying the measuring and weighing systems in the two states. In 221 B.C., Qin Shi Huang issued his imperial edict on the unification of the measuring and weighing systems and required that the edict be carved at the bottom of each apparatus. An apparatus, with descriptions telling the stories of the State of Qin lasting a few hundred years, carries a significance that is much greater than the apparatus itself. Qin Shi Huang was credited for unifying the measuring and weighing systems. Over largely spread areas of land in China, weights and weighing apparatus made in the Qin Dynasty have been unearthed in great quantities. Based on rough statistics, the areas cover all the states unified by the State of Qin. All these revealed the vision and determination of Qin Shi Huang in unifying the measuring and weighing systems in his unified China.

The metrological technologies in ancient China were kept in the historic records in every dynasty, which allows researchers to compare the historical references with real objects that can be found, study the carved descriptions, measure the actual values, and to simulate and resume their functions, thus demonstrating the glory of the ancient sciences and technologies. During the last years of the Western Han Dynasty, Liu Xin, a scholar of Confucian classics, collected and collated the wisdom of his predecessors, and put forward a theory of setting the standards of the measuring and weighing systems by the length of Huang Zhong (a pitch pipe) and by lining up pieces of corn millet, which was recorded in *Han Shu / Lv Li Zhi* later on. This was the resonation of the technologies of determining lengths by audio frequencies and by lining up corn millets as the medium. Sound and capacity are intangible, while capacity can be determined by apparatus. However, in the ancient times, sound was unable to be kept after being determined. It was definitely a challenge for Liu Xin to actualize the scientific assumption of determining measures and weights with the help of Huang Zhong. After countless experiments, Liu Xin suggested taking millets as the medium, namely, selecting 90 pieces of corn millet of medium size as a natural object, and line them up to reach the length of Huang Zhong (9 cun). One chi was reached by adding on 1 cun. The volume of the pipe was some 1,200 pieces of corn millet, which was about the capacity of 1 yue. And 2 yue was equal to 1 ge (c. 16.2 cubic fen). The sum of 1,200 pieces of corn millet had a rough weight of 12 zhu, and 24 zhu was equal to 1 liang. Thus an array of equations with referencing values was formed that determined the values of the units of measures and weights. From then on, these quantitated equations were used as standards in the following dynasties. The temperament was able to serve to determine the measures and weights, and vice versa. Many scholars in modern times, after examining the actual objects and making theoretical calculations, have verified that the equations are tally with principles in physics and the experiments can be resumed within close tolerances.

In ancient China, there were many outstanding scientists who paid close attention to time reckoning and the measuring and weighing systems throughout their researches and related activities which were inseparable from measuring "time, space, and capacity". In the classics of *Mo Jing* published in the period of the Warring States, lever principles were studied and described, which led to further theoretical analyses of the production of weighing apparatus. However, there existed shortcomings in the descriptions that were short of terms in applied mathematics to depict the interactions in balancing. While annotating *Jiu Zhang Suan Shu*, Liu Hui, a mathematician in the times of the Three Kingdoms, compared the units in "Xin Mang Jia Liang" with those of the lengths and capacities in Wei of the Three Kingdoms, and found out that the values of chi and hu in Wei increased 4.7% and 2.6% respectively. Zu Chongzhi, a mathematician of the Southern Dynasty, after countless calculations, reached the value of Pi more accurately with 7 digits after the decimal point (3.1415926-3.1415927). Further more, he cross-checked the data in "Xin Mang Jia Liang" and pointed out that there existed careless omissions in the data presented by Liu Xin. In the Tang Dynasty, after applications for

1,600 years, such units as liang, zhu, lei, and shu in the weighing systems were converted to the decimal systems, thus making 10 qian equal to 1 liang. After further improvement in the Song Dynasty, an array of units for smaller values in the decimal systems, namely, "liang, qian, fen, li and hao" was advocated. It is worth mentioning that in the Song Dynasty, government officials joined in astrologers and arithmeticians in examining the existing measuring and weighing apparatus. Take Liu Chenggui as an example, during the early years of the Song Dynasty, he collated very carefully various types of apparatus used for weighing silver in the Imperial Court and invented a delicate lever scale called Deng Zi, and produced a new set of weights for it. Li Zhao invented Yue Cheng which was also called Water Steelyard. He advocated a standard of taking 1 sheng of water as 1 jin. It took Sima Guang and Fan Zhen some 30 years to debate on whether "measures (were produced) after temperament" or "temperament (was produced) after measures". In the Northern Song Dynasty, a scientist named Shen Kuo wrote *Fu Lou Yi* (a kind of clepsydra) that recorded the structures of the apparatus (Lou Ke) and methods applied to diminish instrumental errors. In *Song Shi – Yu Fu Zhi*, the internal structures, including the gear systems and the machinery mechanisms, of the Compass Chariot (Zhi Nan Che, literally South-Pointing Vehicle) and the Drum Chariot (Ji Li Gu Che, literally Mileage-Reckoning Vehicle) were depicted in details.

Researches on the history of metrology began in the early years of the 20th century. Such authoritative scholars as Luo Zhenyu, Wang Guowei, Liu Fu, Ma Heng, Rong Geng, Shang Chengzuo, Tang Lan, Zhu Dexi and so on published a good number of pertinent works which examined Shang Yang Fang Sheng, weights and weighing apparatus in the Qin Dynasty, Xin Mang Jia Liang, and measures in the past dynasties. In 1937, Wu Chengluo published *The History of Chinese Measures and Weights*, which was the first monograph on the comprehensive history of measures and weights in China. Along with the development of the archaeological excavation since the middle of the 20th century, more and more ancient apparatus for measures and weights were unearthed, which was followed by numerous publications of research papers written by experts on history, cultural relics, archaeology, and paleography. Towards the end of 1970s, a team was built under the leadership of the National Bureau of Metrology with a mission of conducting researches on the history of measures and weights, which resulted in publications on research papers and books in this area. Gradually a branch appeared in the research of the history of science and technology and caught attention from other countries. Owing to the fact that the Chinese language is difficult to learn and understand, there have been lots of requests by foreign friends and researchers to get hold of a book in English on the history of Chinese measures and weights, and to provide them with some materials as references. Hence I had an idea of compiling a book with illustrations to be entitled *The History of Ancient Chinese Measures and Weights*. After years of endeavors, and with the support from the National Bureau of Quality Inspection, and from the Section of Metrology, I was able to present a draft of such a book. Thanks to Mr. Li Shengbing, Managing Director of Beijing Traditional Cultural Development Center, and publishers of Anhui Science and Technology Publishing House, who appreciated and valued the works, the publication of this book became possible.

It is my hope that the book will draw closer attention from readers on the history of Chinese measures and weights, and will raise greater interests from foreign friends in researches on the history of metrology. I also hope that this book can serve as an instrument to promote exchanges of perspectives and approaches in this area. I expect to see a book on comparative metrologies of the world in the near future.

Qiu Guangming

目录

中国古代计量史

THE HISTORY OF ANCIENT CHINESE MEASURES AND WEIGHTS

CONTENTS

CONTENTS

CONTENTS

THE HISTORY  OF ANCIENT CHINESE

EASURES AND WEIGHTS

THE HISTORY OF ANCIENT CHINESE MEASURES AND WEIGHTS

中国古代计量史

一、原始的数量和时间概念

Chapter One

The Primitive Concepts About Quantity and Time

中国计量史的特点，在于它的独立性和连续性，很少受到外来的影响。清朝末年到中华民国(1875～1949)，为了与世界接轨，政府开始改革传统的制度，提出采用米制。经过不断的努力，最终在原有的基础上，采用最简单的换算而达到了逐步向米制平稳过渡。

这里说的计量，包括度量衡在内的其他测量，如时间、湿度、温度、风向、风速等。在中国古代，度量衡专指长度、容量、质量(重量)三个量。古代度量衡，其单位制基本上沿袭了秦代(前221～前207)所确定下来的制度，又经过汉代完备地记于史籍。几千年来，尽管朝代更迭，但总是在继承的基础上求改进。汉代的长度单位是分、寸、尺、丈、引。1引＝10丈＝100尺＝1 000寸＝10 000分。容量单位是龠、合、升、斗、斛。1斛＝10斗＝100升＝1000合＝2000龠。重量单位是黍、累、铢、两、斤、钧、石。1石＝4钧，1钧＝30斤，1斤＝16两，1两＝24铢＝240累＝2400黍。

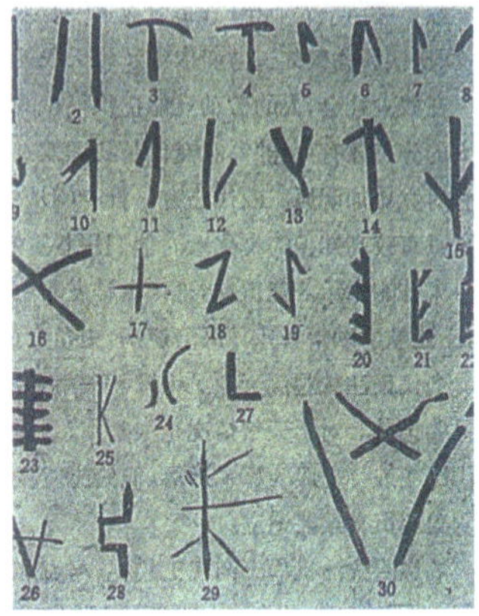

新石器时期陶器残片上常见有记数的符号
Recurring symbols of numbers on pottery remnants in the New Stone Age

A longitudinal view of the Chinese measures and weights reveals their conspicuous features of independence and continuity in their historical development. In Chinese history as long as several thousand years, measuring and weighing systems have developed free from external interference. During the last years of the Qing Dynasty from 1875 to 1949, the traditional Chinese systems began to undergo a process of transformation with the goal of adapting the metric system. This process was basically completed in the Republic of China (1912-1949), when a set of methods of conversion was designed and widely accepted.

In this book, “measures and weights” means not only the measuring of length, capacity and mass (in ancient China, there was no differentiation between mass and weight), but also that of time, humidity, temperature, wind direction, wind speed and so forth.

Units and their fractions in the measuring and weighing systems were designated and set in the Qin Dynasty (秦代, 221B.C.-207B.C.), and recorded systematically in books in the Han Dynasty (汉代). Throughout the history ever since, with dynasties changed, units and their fractions set in the Qin Dynasty and perfected in the Han Dynasty were basically followed, although there existed improvements. In the Han Dynasty, units and their fractions in the linear measures comprised fen (分), cun (寸), chi (尺), zhang (丈), and yin (引). 1 yin =10 zhang, 1 zhang =10 chi, 1 chi = 10 cun, 1 cun = 10 fen. For the measures of capacity there were yue (龠), ge (合), sheng (升), dou (斗), and hu (斛). 1 hu =10 dou, 1 dou = 10 sheng, 1 sheng = 10 ge, 1 ge =2 yue. For the measures of weights there were shu (黍), lei (累), zhu (铢), liang (两), jin (斤), jun (钧), and dan (石). 1 dan = 4 jun, 1 jun = 30 jin, 1 jin = 16 liang, 1 liang = 24 zhu, 1 zhu = 10 lei, 1 lei = 10 shu.

新石器时期的七孔石刀，孔距均匀，孔眼相当，是经过比较测量的
An unearthed seven-hole stone chopper in the New Stone Age, with the holes regularly distributed, showing evidently the application of measurement

母系氏族社会村落遗址复原图。房屋大小适当，排列有序

Shelters, with appropriate sizes and regular arrangement, erected in a village ruins restored from the matrilineal commune times

计量活动是用数值来表示事物的量。因此最早的测量是在学会记数之后才开始的。原始人群通过劳动、生活和分配，对数和量逐步有所认识，并产生了表示数量的方法，如结绳记事和按量估堆等。人类对长度的测量，大约是从天然洞穴里搬迁出来，靠自己的双手建造房屋时开始的。在陕西省发掘出多处母系氏族社会村落遗址，其中半坡是典型的一处，分居住区、手工业区和公共墓地三部分。居住区房屋式样有方形和圆形，经实地观测，设想当时人们在动工前已有了布局规划，并确定朝向、间距以及测量每座房屋的面积等。其中一间大房子可能是公共活动场所，它每一边的长度约略相等，是当时经过测量的实证。

Measures and weights have been used essentially as numerical values to represent quantities of objects. So the measures and weights came into being after people had learned the way of counting. Primitive people became aware of numerical and quantitative values through their experiences of labor and life, thus starting to count numbers and estimate quantities in very simple ways, for example, to tie knots in a cord to count numbers and to pile objects to estimate quantities. After they moved out of natural caves and began to build shelters, people learned gradually the method of linear measures. Relics of villages inhabited by the people then showed matriarchal commune patterns, which were unearthed in Shaanxi Province. They were typical ones of those villages that consisted of the residential quarters, handicraft quarters, and cemeteries. Seeing from the layout of houses in a residential area, we realize that before the construction started, people had made the plan in which measures must play an important role. In the residential quarters there was a big house that was probably used as a public building. The fact that the four sides of the big house were almost of the same length can be served as a proof that the concepts of measurements were applied in practice then.

新石器时期陶器残片上已有排列规整的几何图形

There existed geometric figures spread evenly on pottery remnants in the New Stone Age

父系氏族社会遗址出土了数件残陶器，经测量，它们之间的容积有一定的倍比关系，可能是测量器具
Several pieces of unearthed pottery on the ruins in the patriarchal society, with certain relationships of times in capacity, supposedly used as measuring instruments

A.大地湾 901 号房遗址
No. 901 Ruins in Dadiwan

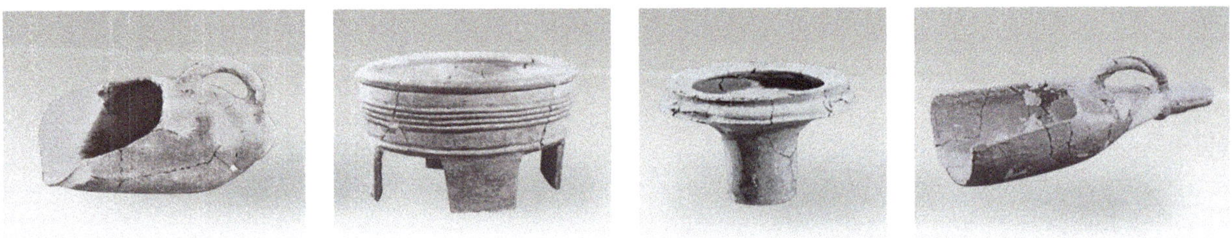

B.原始陶量复原图
Primitive pottery measuring instruments restored

父系氏族社会时期彩陶上刻有表示日、月、山、树的符号
Symbols of the sun, the moon, mountains and trees carved on colored pottery in the patriarchal society

寒与暑是很容易感受到的自然现象，人们从通过感觉器官来判断季节的变化，到学会用数来记录从月圆到月缺所经过的 30 个昼夜，进而发现太阳和星星移动的位置规律等，逐步形成了最初的天文历法知识，即年岁、季节和物候月，开始了种植活动。这些天文现象，可以从出土的原始社会陶器残片上绘制的日、月、山，日出、日没以及把太阳在星空背景上绕一周的路径均匀地分成 12 个等分的图案上得知。据文献记载，氏族社会后期，传说中的氏族领袖颛顼，已通过观测星辰推算一年的长度。黄帝则创立了度、量、衡、里、数五个量，又命大臣尧创造了记日的方法。

继黄帝之后是尧、舜。尧命羲和根据日月星辰的运动规律来制定历法，把一年定为 366 日。由于一年的长度不是一个月的长度的整数倍，这时期的历法已采用了置闰月以调整。舜时又对各部族的历法和度量衡作了协调统一。

A. 大河村遗址
Ruins of Dahe Village

B. 太阳晕珥图
The solar halo

C. 太阳图
Solar images

D. 星象图
Star images

E. 太阳图像的岩画
Solar images on cliff drawings

母系氏族社会遗址中出土的有关天象纹饰和图案的陶片
Pottery pieces with astronomical maps, grains and patterns unearthed in the matrilineal society

Measures of time also originated in the primitive times in China. In remote antiquity, Chinese people had observed that the period from the crescent to the full moon was 30 days, and had learned the way of recording different time by using figures in this period. Further more, they discovered regularities of the movement of planets and the moon, and started to measure time by using these regularities. Such discoveries from observation of astronomical phenomena gradually formed up the earliest concepts of year, season and month, and this played a role of guiding agricultural activities. On some potsherd there are patterns of the sun, the moon, sunrise and sunset. On some other potsherd the orbit of the sun was drawn and divided into 12 parts evenly. It was recorded in Chinese historical documents that in the late times of the clan society a legendary man named Zhuan Xu (颛顼), who was the chief of a tribe, began to calculate the length of a year by observing the changing positions of planets and stars. Huang Di (黄帝) established the system of measuring and weighing, and instructed one of his ministers named Yao (尧) to create the way of calculating the length of a day.

Huang Di's successors Yao and Shun (舜) also contributed a lot to the primary measuring and weighing systems in their times. Yao ordered Xi He (羲和) to formulate a calendar, following the regularities of movements of the stars, the sun and the moon. According to this calendar, there were 366 days in a year, and a leap month was set in order to make necessary adjustment. Shun unified calendars and the systems of weighing and measuring in different tribes.

THE HISTORY OF ANCIENT CHINE

MEASURES AND WEIGHTS

2

THE HISTORY OF ANCIENT CHINESE MEASURES AND WEIGHTS

中国古代计量史

二、夏商周时期度量衡器具的出现

Chapter Two

The Earliest Measuring and Weighing Apparatus in the Xia, Shang and Zhou Dynasties

继舜之后是禹。为了治理水患，禹创造了测量工具——规(测圆)、矩(测方)和准绳(测长)，并以自己的身长、体重作为长度和重量的标准。禹治水有功，被舜立为继承人，于公元前约2070年建立了第一个王朝——夏。

中国古籍中有“布手知尺”的记载。今传世的几支商代(前1600～前1046)骨尺，其长度在16～17厘米，正合中等身高者伸开拇指与食指之间的距离。尺上划分成10等分，即“寸”。有的尺在寸格中更细分10个等分，即“分”。

西周时期(前1046～前771)人们已将田亩划分成“井”字形，由国家将每方里的土地按“井”字分成九份，中间一份为公田，其余八份分授给八家耕种，公田则由八家助耕，全部收获均交给国家。由此可知，这时期已有“里”和“田”这样的面积单位了。西周时期的青铜器上还刻有重量单位“寽”和“匀”。王室和领主常常把一定量的铜作为赏赐和交换的等价物。证明这时期的重量单位已在上层社会普遍应用。为满足粮食的积累和计算的需要，容量单位也开始出现，有稷、秉、缶等。商周时期度量衡器具往往都藏于王府，王室和领主利用这些器具来收取赋税或为上层社会的某些需要服务。

史书记载，父系氏族社会时期，黄帝已提出设立度、量、衡、里、数等五个量。少昊建议“同度量”

According to historical records, as early as in the patriarchal times, Huang Di advocated the implementation of measuring and weighing and such units of mu and li, and Shao Hao suggested “Unification in Measuring and Weighing”

In Chinese history, the Xia Dynasty (c. 2070B.C.-1600B.C.) represented the first slave society. Its first ruler Yu (禹), who was the successor of Shun (舜), invited many types of measuring apparatus for water control. These apparatus included gui (规) (gauges for the measuring of round objects), ju (矩) (for the measuring of square objects), and zhunsheng (准绳) (for the measuring of length). Yu was also using the height and weight of his own figure as standards in measuring and weighing. Credited for his accomplishments, and upheld by Shun, Yu founded Xia (夏), the first dynasty. There were records about some popular methods of liner measurements, such as chi (尺) in Chinese ancient books— the length of chi was obtained by measuring the hand (布手知尺). The length of several unearthed bone rods made in the Shang Dynasty(商代,1600B.C.-1046B.C.), which is 16-17cm. The length is just equal to the span of a man of medium height. Carved on those bone rods are 10 subdivisions, which are called cun (寸), and 1 cun can be further divided in 10 fen (分).

During the times of the Western Zhou (西周 1046B.C.-771B.C.) the jing-shaped (井) measuring methods were adopted in the measurements of tian (田) and mu (亩). The then government divided field of 1 square li (里) into 9 fractions evenly, which looked like the Chinese character “井” (in English literally, “well”). In this divisional structure of fields, the part at the center was communal with the harvest for the state, and the 8 parts around the center were assigned to 8 families. The existence of this system shows that the units such as li and tian (田) were used in the measurement of area. On some bronze wares made in the Western Zhou times unit of

weight such as lue (寽 lüè) and yun (匀) were engraved. At that time, bronze was often used as rewards by the king, and occasionally taken as equivalence in trade. All these are the proof that in the Western Zhou times units of weights came into being, and were used at least in the upper-class society. Meanwhile, in order to calculate the accumulation of grain, which was an inevitable outcome of the development of productivity, units such as zong (稯), bing (秉) and fou (缶) and so forth were used for measuring capacity, and measuring apparatus were invented. During the time of the Shang and Zhou dynasties, those apparatus were usually kept in the royal court for the convenience of the king and his ministers to ensure the income of the state treasury by levying taxes, or to meet some special needs.

唐代彩绘伏羲、女娲手执规矩图
Fu Xi and Nu Wa, holding compasses and rules on colored drawings in the Tang Dynasty

汉代画像砖刻伏羲、女娲手执规矩量度天地
Fu Xi and Nu Wa, holding compasses and rules to measure heaven and earth, on brick drawings in the Han Dynasty

商周时期已建造大型宫殿，应有各种测量工具
Various types of measuring instruments were used in building large-scale palaces in the Shang and Zhou dynasties

《史记》记：大禹治水“左准绳、右规矩”。用“规”画圆、“矩”测量角度，“准绳”是有一定标准的测长工具
As recorded in *Shi Ji*, Da Yu, while regulating rivers, used zhunsheng, compasses and rules as standardized measuring instruments for drawing circles and measuring angles

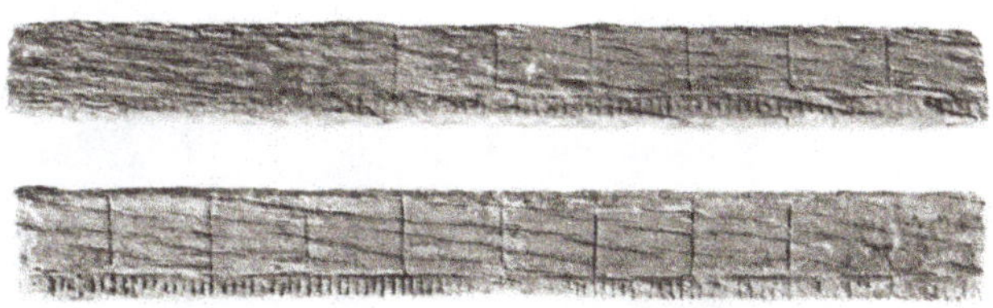

两支商代骨尺，长 16 ～ 17 厘米。约略相当于中等身高人拇指与食指伸开的距离，与文献中“布手知尺”和象形文字“尺”形象相合
Two bone rulers in the Shang Dynasty, 16-17cm in length each, approximately the length of a span, which matched the “Measuring the Span for Chi ” and the pictographic character of chi

西周时期 铜卣上有土地面积"方五十里"的铭刻

A bronze vessel in the Western Zhou Dynasty with inscriptions indicating areas of land, "50 Square Li"

商代 父己铜觯上刻有"疆"字，写作"畕"，两个"田"字中间有"▽"形符号，表示用矩测量土地面积和定疆界

A Fu Ji bronze goblet in the Shang Dynasty with inscriptions and symbols of triangles, indicating the implementation of rules in measuring field areas and in determining borders

西周时期 不娸铜簋铭文记周宣王赏给不娸"十田"等

A round-mouthed bronze vessel in the Western Zhou Dynasty with inscriptions of "10 Tian" awarded to Bu Ji by the king Zhou Xuan Wang

西周时期 伯格铜簋 铭文记伯格用四匹马与倗生交换了“卅田”。双方还实地勘定了“田界”等

A round-mouthed bronze vessel in the Western Zhou Dynasty with inscriptions of “30 Tian” for which Bo Ge exchanged 4 of his horses with Peng Sheng

西周时期 铜匜 全篇铭文是一篇诉讼判决书，其中有施于受罚者鞭刑五百、罚铜三百寽等内容

A gourd-shaped ladle in the Western Zhou Dynasty with inscriptions of lawsuit and judgment in which such court decisions were stated as punishment of caning 500 times and a fine of 300 lue, etc

西周时期 禽簋 铭刻中有“金百寽”，“寽”是重量单位

A vessel in the Western Zhou Dynasty with inscriptions of “Jin Bai Lue” —Lue was a unit of weight

战国时期 房屋(明器)
Houses in the Warring States (in the nether world)

战国时期 中山王陵园建筑平面图 均以尺、寸为单位标出各种建筑的大小、间距
Maps of the Zhongshan Wang Tombs in the Warring States with indications of chi and cun for measuring the sizes and spaces

公元前约2400年,古埃及人在压缩草板上所绘征收赋税时牵绳测量、划定地界图
Ancient Egyptians, in c. 2400 B.C., were drawing on a compressed grass board ropes for measuring in levying taxes and in dividing fields

古埃及牵准绳划定建筑物地基
Ancient Egyptians were measuring the buildings and grounds with zhun sheng (ropes)

战国时期 铜矩尺 各边皆长 23.1 厘米,与秦、汉尺同长
Bronze rulers in the Warring States with a length of 23.1 cm on each side, the same length as the rulers used in the Qin and Han dynasties

春秋时期 矩形建筑铜构件
Square bronze structural parts used in the Spring-Autumn Periods

THE HISTORY

OF ANCIENT CHINES

EASURES AND WEIGHTS

3

THE HISTORY OF ANCIENT CHINESE MEASURES AND WEIGHTS

中国古代计量史

三、先秦时期历法的制定、时间的测量

Chapter Three

The Calendar and the Measuring of Time in Pre-Qin Times

什么是历法？简单地说，就是人们为了社会生产实践的需要而创立的长时间的记时系统，也就是对年、月、日、时的安排。

Calendar, put in simple words, is a system of time-keeping in which year, month, day, and hour were scheduled. Hence calendar is used to meet the needs of various social practices.

《隋书》记：周公测晷影于阳城，以考历记
In *Sui Shu*, it was recorded that Zhou Gong measured the shadow cast by the sun on a sundial in Yang Cheng, in order to cross check the calendar

周公在阳城（今河南告城）设台观测日影。这是中国最早、保存最好的古天文遗迹
Zhou Gong observed the shadow cast by the sun on a sundial in Yang Cheng (presently Gao Cheng, Henan Province), the earliest and most intact site for astronomical observation in ancient China

1. 先秦时期历法的制定

现存最早的一部物候历，是成书于春秋时期的《夏小正》，它所记载的时日制度是夏代流传下来的。其中按 12 个月，逐月记录了当时的物候、天象、农事活动和重要的政事。不仅反映出夏代对物候观察的内容已很丰富，而且把物候知识用来为农业服务了。

商周时期，大规模的祭祀和占卜要求有准确的祭祀周期。加上农业的进步，社会生活更高的需求，使商代的历法在夏代的基础上有了进一步发展。商周时期的历法为阴阳历，阳历以地球绕太阳一周，即 365.25 日为一回归年，又称"四分历"。阴历以月亮绕地球一周为一朔望月。商周用干支纪日，数字纪月，月有大小之分，大月 30 日，小月 29 日。12 个朔望月为一个民用历年，它与回归年有差数，所以阴历在若干年内置闰月。季节与月份已有大体固定的关系，商周卜辞中常见"春""秋"之称。一天之内又分若干个时间段，如旦、朝、中日、暮等。"旦"为日初之时，"朝"为天明，"中日"为中午，"暮"为日将落之时。

春秋战国时期天文历法知识与商周时期相比，最大的特点是已从一般的观察向数量化观察过渡。这时期的文献如《左传》《春秋》《礼记》《吕氏春秋》等都有丰富的天文知识的记录。《左传》中就记载了鲁国的国公僖亲自参加了制定天文历法的活动，以及通过测量日影这类较科学的方法来划分四季，定二分（春分、秋分）、二至（冬至、夏至）以及立春、立夏等节气。

1. The Formation of Calendar in the Pre-Qin Times

The oldest phenological calendar in ancient China is known as *Xia Xiao Zheng* (《夏小正》), which was compiled in the Spring and Autumn Period (春秋时期). The arrangement of time in this book was inherited from the Xia Dynasty (夏代). In this book a year was divided into 12 months, and phenology, celestial phenomena, agricultural practice, and important political events in each month were recorded. This book itself indicates that in the Xia Dynasty the content of phenology was quite rich, and was very useful in agriculture.

During the times of the Shang and Zhou dynasties (商代和周代), the sacrificial rites gained an important position in the social life, which required a more precise calculating of the sacrificial time and periods. In addition, the progress of agriculture made the improvement of calendar a must. Therefore, the calendrical science was developed in the Shang Dynasty on the base of the achievement in the Xia Dynasty. In the periods of Shang and Zhou, the calendar was a lunar-solar system. In the solar calendar, the time of the earth's revolving around the sun, i.e., 365 and 1/4 days, was a tropical year. That is why the solar calendar was also called "four-divisional calendar". In the lunar calendar, the time of the moon's revolving around the earth, that is, 29 or 30 days, was a synodic month. At that time, calendrical scientists attempted to arrange short and long months so that the moon's conjunction would take place on the first day of every month. The day donation was by using gan zhi (干支, Heavenly Stems and Earthly Branches), while the month donation was by using figures. In the lunar-solar calendrical system, in order to patch up the discrepancy between the lunar year and the solar year, an intercalary month was set in every few years. In this system, a day was divided into several sections of time, which was donated as dan (旦) for the time of sunrise as the beginning of the day, zhao (朝) for the time of sunrise, zhongri (中日) for the time of noon, and mu (暮) for the time of sunset. Also in this system, the relationships between seasons and months were roughly fixed.

In comparison between the Shang and Zhou dynasties, the calendrical science in Spring and Autumn Periods and in the Warring States Periods bore an obvious characteristic, namely, figures were introduced in the observation of astronomic phenomena. Documents of these historical periods such as *Zuo Zhuan* (《左传》), *Chun Qiu* (《春秋》), *Li Ji* (《礼记》), and *Lv Shi Chun Qiu* (《吕氏春秋》) contain many records of calendrical science. For example, in *Zuo Zhuan*, it was recorded that the king of the State of Lu (鲁国) named Xi (僖) personally participated in the making of calendar. It was also recorded in this book that the division of a year into 4 seasons, the fixing of the Chun Fen (春分, Vernal Equinox) and Qiu Fen (秋分, Autumnal Equinox) as well as that of Li Chun (立春, Beginning of Spring) and Li Xia (立夏, Beginning of Summer) could be done by measuring variations of the shadow of the sun.

古罗马的日晷
A sundial in ancient Rome

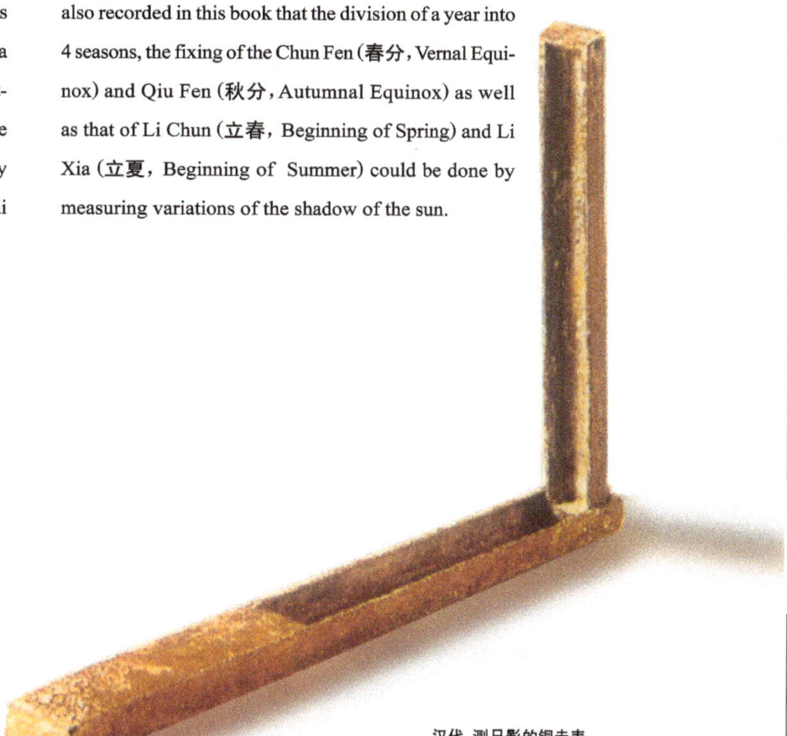

汉代 测日影的铜圭表
A bronze sundial in the Han Dynasty to measure the length of the shadow projected by the sunlight

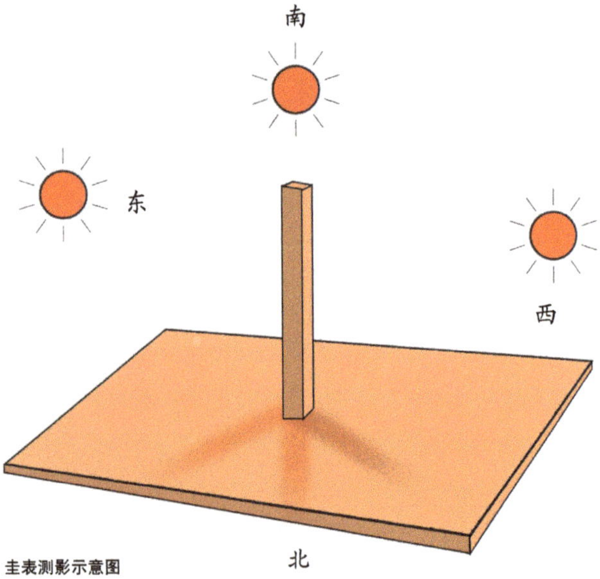

圭表测影示意图
An illustration of how the sun was casting shadow on the sundial

图间插文 Descriptions matching the Pictures

中国古代计时方法是与天文测量分不开的，年、月、日、时都离不开对天象的观测。

In ancient China, time reckoning was implemented together with astronomical surveying, thus year, month, day and hour could not be separated from astronomical observation.

2. 先秦时期时间的测量

时间计量单位和长度、重量一样，是人为规定的。但是实践证明，利用与农业生产关系密切的自然现象变化规律，作为天然的时间单位的尺度，对计量时间将会带来极大的便利。因此，时间计量与历法相结合，便形成了一套自然的计时单位。如“日周于天，一寒一暑，四时备成……谓之岁”，四时即四季。“月一盈亏谓之月”，“凡日一出没谓之一日”。日常生活中，人们接触最多的自然时间单位是“日”，但以日计时远不能满足人们的需要，于是便出现了将一日分成 10 时、12 时和 15 时。中国古代普遍采用的是把一日分成 12 个时辰的计时制度。当人们需要对时间做更细、更精确的划分时，又有了把昼夜分成 100 刻制(1 刻约为 14.4 分钟)。百刻制完全不考虑太阳运动的规律，是一种人为的时间单位，因此不便于在天文学上使用，而 12 时辰制比较符合天文学上的习惯，却划分过粗。这样，两种制度就难以相互取代而长期共存了。

漏刻是古代重要的计时工具，早在史前时期开始萌生，至西周时期已成为制度，并设有专门管理漏刻的官吏——挈壶氏。漏刻以百刻分昼夜，冬至昼漏 40 刻、夜漏 60 刻，夏至昼漏 60 刻、夜漏 40 刻，春秋各 50 刻。

漏刻与圭表计时相结合，直接促成了日用计时和科学计时体制的建立。圭表是最古老、结构最简单的一种天文仪器。古人很早就观察到，直立的标杆影长不断地随着太阳的位置不同而变化。观察日影方位的变化，就可以逆推太阳的空间方位，从而得知相应的时刻。用杆影测定中午的时刻其精度很高，是古代用来校准漏刻记时的主要方法。这种想法的实施，导致了日晷计时的诞生。

刻有干支、日蚀的甲骨。干支用以记年和日
A tortoise shell with celestial stems and earthly branches carved on—for recording dates

2. The Measuring of Time in the Pre-Qin Times

Units in the measurement of time, just like that in the measurement of lengths and weights, is artificially set. However, it was proved time and again that taking the law of the change in natural phenomenon as the base for setting units of time is reckoning of great convenience. Therefore, in ancient China time reckoning was integrated into almanac, thus a set of natural units of time reckoning was invented. In historical documents it was recorded that "the cycling of the sun leads to the alternation of the cold season and the hot season, based on which four seasons are reached… These four seasons form up a year." "A month begins with the observation of the first visible crescent and ends with the last." "A day begins with the sunrise and ends at the sunset." In daily life "day" was the most popular used unit of time-keeping. But this unit was far from enough for time reckoning in daily life, so subdivisions occurred. In ancient China, one of the subdivisions was shi (时 close to the concept of hour). A day could be further divided into 10, 12, or 15 shi, among which the second one was most frequently used. When people needed to calculate time more precisely, the subdivision of ke (刻 close to the concept of a quarter of an hour) was created, and a day was divided into 100 ke. Then 1 ke is equal to 14.4 minutes. Ke as a unit of time reckoning has nothing to do with the moving of the sun, so it could not be used in astronomy. Shi somehow coincided with the changing of astronomical phenomena, but this unit was too "big" to calculate time precisely. That is why shi and ke had co-existed for a long time in the Chinese history. As one of the important time-keeping instruments, clepsydras (漏刻) were originated in prehistoric ages. In the time of the Western Zhou (西周), clepsydras became an important instrument for ensuring precise time keeping, and the government even assigned officials called Qie Hu Shi (挈壶氏) to manage clepsydras. On a clepsydra there was 100 graduations indicating day and night. In winter, the daytime would read 40 graduations and the night 60 graduations. In summer, the daytime would read 60 graduations and the night 40 graduations. In the spring and autumn, the daytime and the night would equally read 50 graduations.

Parallel to the application of clepsydras was the use of Gui Biao (圭表), an ancient Chinese time keeping instrument. It was the combination of applications of the two types of instruments that laid the corner stone of the scientific systems of time keeping. Gui Biao, simplest in structure, consisting of an elongated dial (圭, gui) and 1 or 2 gnomons (表, biao), was actually the earliest time keeping instrument in ancient China. In remote antiquity, Chinese people had observed that the length of the shadow of an upright pole would change with the movement of the sun, so the position of the sun can be known by observing the change of the shadow, and the time can thus be calculated. Since determining the time of the noon by observing the shadow of an upright pole is of high precision, ancient Chinese people then made adjustments accordingly on clepsydras, which resulted in the invention of Ri Gui (日晷, sundial).

《尚书》记：西周设掌管、发布时间的各级官吏
In *Shang Shu*, a complete system of levels of officials was named to be in charge of and announcing the time

《左传》记：鲁国的国公僖(前669~前626)亲自主持测日影以划定四季
As recorded in *Zuo Zhuan*, Lord Xi of Lu (669 B.C.-626 B.C.) personally took charge of measuring the shadow of the sun for dividing the four seasons

THE HISTORY

OF ANCIENT CHINESE

四、春秋战国——度量衡发展的繁荣与混乱时期

Chapter Four

The Spring and Autumn and Warring States Periods: Prosperity and Chaos of the Measuring and Weighing Systems

西周末年，奴隶制经济逐步崩溃，社会各方面矛盾激化。公元前771年都城被攻破，周幽王被杀。太子平王继位后不久把都城迁到地处东部的洛阳，史学家以平王东迁为界将周分为西周和东周。东周是一个名存实亡的王朝，分布在各地的大大小小的邦国，都趁王室衰败之机积极地发展各自的势力，企图取代周王室的地位，导致战乱不断。故史学家又把这一时期称为春秋战国时期(前770～前221)。

春秋战国时期在中华文明发展史中是一个十分重要的历史阶段。一方面列强称雄争霸，战乱此起彼伏；另一方面各民族、各地区的文化却得到空前的交流、融合，社会处在大变革、大动乱之中。又由于百家蜂起、诸子争鸣而形成了开拓、创新的氛围，出现了一个前所未有的繁花似锦的时代。度量衡的发展繁荣和混乱，正是当时社会政治、经济发展的一个侧面的反映。

东周时期(春秋)图。周王室东迁后，形成了诸侯国割据争霸的局面

A map of the Eastern Zhou (the Spring and Autumn Period) times

During the last stages of the Western Zhou (西周), the slave system began to fall apart. In 771B.C., You Wang (幽王), the last king of Western Zhou, was killed in a riot. The prince, who was enthroned as Ping Wang (平王), moved the capital of Zhou eastwards to Luoyang (洛阳). From then on, the Zhou regime was known as the Eastern Zhou (东周). The era of Eastern Zhou was divided into two periods: the Spring and Autumn Period (春秋) and the Warring States Period (战国). (770B.C.-211B.C.) Towards the end of the Eastern Zhou, the power and prestige of the royal house was gradually declining, while individual states led by their war lords grew in status and influences. When each of those states wanted to topple down the throne, wars broke out frequently. That is why this period of time in Chinese history was called the Warring States Period.

The Spring and Autumn and Warring States Periods is an important stage which witnessed the development of the Chinese civilization. On the one hand, individual states scrambled for supremacy, thus making situations chaotic and unpredictable. On the other hand, in such social and political upheavals, cultures of various nationalities and of different areas were interwoven and integrated. In addition, many schools of philosophies flourished, with each presenting its own teaching, propagating its own beliefs while criticizing those of others. Together they created a situation in which "a hundred schools contended". This situation was created with an atmosphere that encouraged innovation. Synchronizing with the social and political changes and chaos was the development and prosperity of the weighing and measuring system.

秦

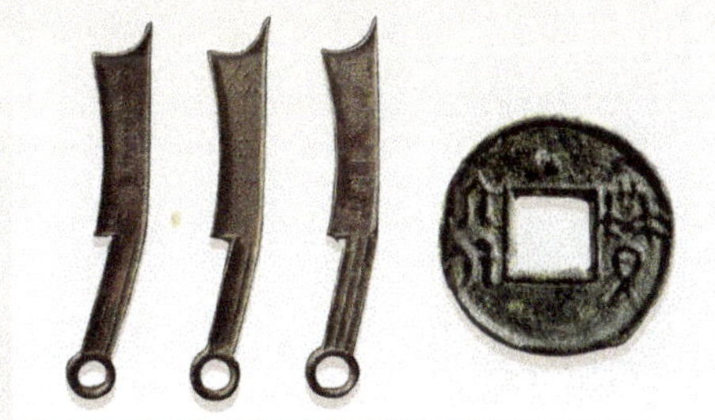

齐

三晋(韩、赵、魏)

楚

各诸侯国为发展经济而大量铸造货币。各国的货币形式不同，重量单位各异

In each of the separatist regimes, for their economic development, currencies were cast with various shapes and units of weight

1. 新的生产关系促进了度量衡的发展

春秋战国时期是中国封建社会初步建立的时期。政治体制的巨大变化，直接影响着经济的变革。土地国有制逐步被封建领主私有制取代，部分公社农民转化为小土地所有者。土地占有形式的改变使井田制受到冲击并走向瓦解，劳役地租开始向实物地租转化，新的生产关系促进了生产力的发展。粮食增产了，其中由农民支配并转化为商品的数量也随之增加。农业生产关系的改变促进了手工业和商业的发展，出现了官府手工业之外的个体手工业和农民家庭手工业，他们生产的产品全部或大部要拿到"市"上去卖，扩大了商品经济的范围。私有财产的增加，私有观念的加强，交换时人们对大小、多少、轻重以及所付出的劳动价值开始计较起来，这些都使过去那种以物易物的交换方式无法适应。因此就要有一种公认的、价值稳定、分割简便、坚固耐久又便于携带的商品作为等价物，于是产生了货币。货币的出现，其不可缺少的媒介是度量衡。战国时期铸币现已大量出土，铸币上常常铸有地名和单位面值。尽管受当时铸造水平的限制，有不少面值相同的铸币实际重量参差不齐，但各种货币与重量单位有一定的联系却是必然的。

春秋战国时期仍以官府手工业占主导地位。官府手工业组织法规严谨，分工细致、缜密，有"百工"之称。在手工业制作中对产品质量的检验已提出要求，除要测量长度、重量、容量之外，还要测量角度、温度、硬度等各项指标。迅速发展的手工业，也推动了度量衡的发展与繁荣。

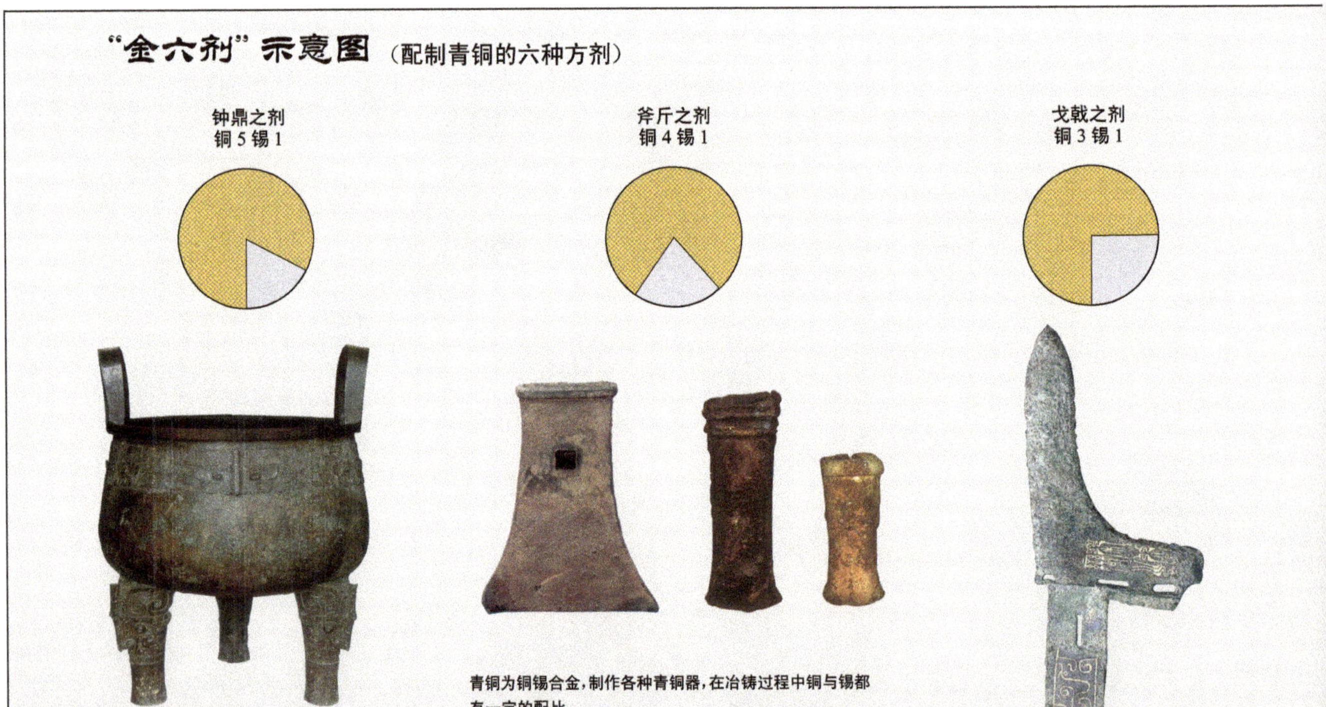

青铜为铜锡合金，制作各种青铜器，在冶铸过程中铜与锡都有一定的配比

An illustration of the "Six Ratios" that set the proportions of copper and tin in the production of various forms of alloy

1. The New Production Structure Promoting the Development of the Measuring and Weighing System

During the time of the Eastern Zhou Period, the feudal system was established gradually. Drastic changes in the political arena exerted a great influence on the economic aspect of social life. In some of the states, the well-field system was abolished, and private ownership of land was legally recognized. Meanwhile, labor hiring was replaced by field leasing. The new production structure and labor relations promoted the development of the productivity. Agricultural areas were expanding continuously. New farm tools and agricultural technology were adapted. Grain yields increased as a result of the quantity of harvest that was owned by peasants who were allowed to trade their grain as commodity. During this period of time, handicraft progressed, and most or all of the products were allowed to be sold in periodical gathering places now known as "markets". The increase of privately owned property in turn stimulated the conceptualization of private ownership, and people started to be cautious and calculating about the volumes and weights of objects in trade. All of these made the "objects for objects" appear inappropriate and inconvenient, thus currency came into being with intrinsic value —public recognition, stable worth, portability and so forth. From the beginning of the use of currency, the measuring and weighing system functioned as an indispensable intermediary. During the Warring States Period, various kinds of coins were forged with different units and weights indicating the differences in value. As the forging technologies developed at different levels in the states, some coins with the same values carried different weights. But it was for sure that values of coins were closely associated with the weights.

In the handicraft industry, workshops and their operation were owned and controlled by the government. In those workshops, rules and regulations were strict, and divisions of labor were elaborated with a term of baigong (百工). Meanwhile, official bodies had set rules for testing the quality of products, which ascertained the measuring of length, weight, angle, temperature, hardness and volume. The development of handicraft industry was an important push factor for the development and prosperity of the measuring and weighing system.

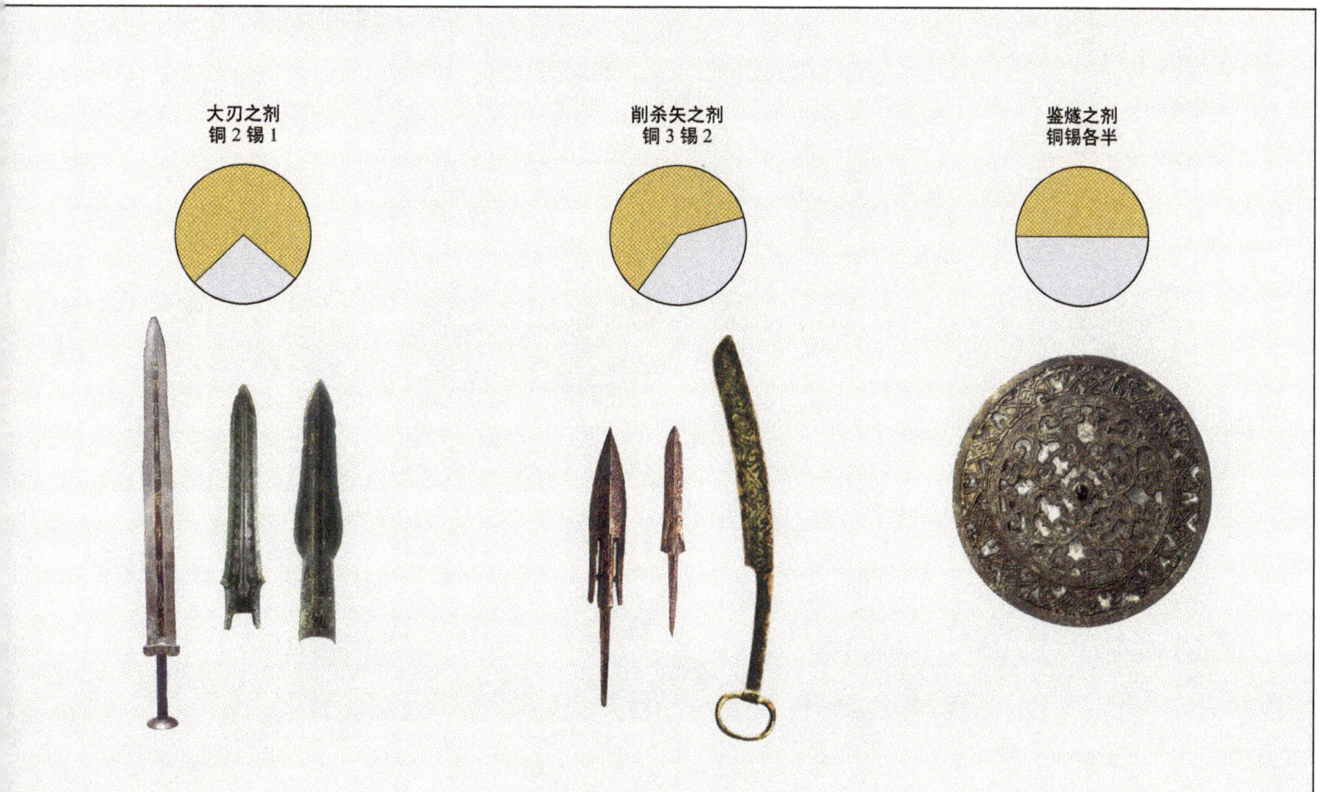

2. 诸侯割据时期度量衡从混乱到逐步趋向统一

春秋战国时期度量衡从建立到逐步健全，又从混乱到趋于统一，是目前所见度量衡器较多的时期，又是度量衡单位制错综复杂的时期。由于诸侯割据，政权分裂，决定了度量衡的不统一。又由于在兼并战争中，各民族、各地区交通的贯通，文化的交流，技术的发展，度量衡也得到交流和发展，并有了统一的要求和愿望。下面以几个大的诸侯国为例，分述这一时期的状况。

2. The Unification Tendency for the Measuring and Weighing System

The Warring States Period witnessed two transitions in the measuring and weighing system: the development was from initiation to amplification, and from confusion to unification. That was the time when numerous types of measuring apparatus were created, thus making the system of unit a great confusion, which was inevitable in the chaotic social and political life. On the other hand, wars provided opportunities for the exchange of different measuring and weighing systems, and during these exchanges demands on unification were growing. The overall situation of the development of the system is illustrated below with the situations in some main states.

春秋时期 铜权 重 212.6 克
A bronze weight in the Spring and Autumn Period, weighing 212.6g

战国时期・齐 右里铜量 容 1024 毫升
You Li Bronze Instrument with a capacity of 1,024 ml in Qi of the Warring States Period

战国时期・齐 右里铜量 分别容 1025 毫升、206 毫升，约相当于 5 升和 1 升量
You Li Bronze Instrument with capacities of 1,025 ml and 206 ml, equivalent to 5 sheng and 1 sheng respectively, in Qi of the Warring States Period

(1) 齐国的度量衡

齐国东临渤海，有渔盐之利，具备了发展经济的天然条件。齐国主张与各地商人相互往来，出现了繁华的商业城市。在这样的国度里，度量衡已是征收赋税、平准物价的日常用器具了，因此对度量衡也提出了较高的要求。《管子》中有"衡石一称，斗斛一量，丈尺一綧制，兵戈一度，书同名，车同轨"的记载。度量衡一旦从殿堂上走向民间，也就必须具备一定的法制性。文献记载齐国的容量单位是"豆、区、釜、钟"：四豆为区，四区为釜，十釜为钟。这是诸侯王制定的制度，称"公量"。后来有权势的士大夫为了与诸侯王相抗衡，又另立"家量"。单位制是：五豆为区，五区为釜，十釜为钟。1857 年出土了三件铜量器，上面都刻有铭文，证明它们是当时齐国置于关卡上，用来测量、校准往来和进出仓廪粮食的标准量器，其中的一件还刻有检定制度和惩处条例。两件自铭为"釜"，实测容量均约为 20000 毫升；一件"鉌"，约容 2000 毫升。近年来还先后出土了数件陶质和铜质的量器，有的陶质量器上刻"升"，1 升约容 200 毫升。几件铜质量器分大、小两种容积，大的约容 1000 毫升，小的约容 200 毫升，正好是 5 倍之比。故知大铜量是 5 升量，小的是 1 升量。那么 1 鉌即相当于 10 升，1 釜相当于 100 升。

经过对不同时期出土器物的实测，又与文献互为佐证，可以认为随着齐国政权的变动，度量衡也在不断变化和改进，从四进位的"公量"，到五进位的"家量"，逐步又走向了十进制。

在各地博物馆还收藏有齐国的铜权，制作十分精美。从文献记载可知，齐国重量单位有"镒"。24 两为 1 镒，从刻有"镒"的器物上得知，1镒约合今 370 克，1 斤约为 246 克。

A.正置
Upright

B.倒置
Upside down

战国时期·齐 有粉(升)字戳印的陶量
Pottery weighing instruments, in Qi of the Warring States Period

战国时期·齐 右伯君铜权 重 198.4 克
You Bo Jun Bronze Weight, 198.4 g in weight, in Qi of the Warring States Period

图间插文 Descriptions matching the Pictures

春秋战国时期，商品经济活跃，手工业发达。青铜器的铸造、竹木器具的制作都离不开度量衡。

During the Spring and Autumn and Warring States Periods, commercial activities were very lively and handicrafts were advanced. Brass founding and bamboo articles would be impossible without the application of the measuring and weighing system.

(1) The Measuring and Weighing System in the State of Qi (齐国)

Facing the Bohai Sea (渤海) in the east, the state enjoyed exceptional advantages of developing fishing and salt-making. The economy of this state was quite prosperous. Just like any state that was economically prosperous, the measuring and weighing apparatus were commonly used in tax-levying and price controlling, thus the unification of the measuring and weighing system was crucial. It was recorded in *Guan Zi* (《管子》) a clear description of request on the unification of units for measuring and weighing, no matter they be in the military, publication and transportation. Once the measuring and weighing system became a necessity in daily life, there must exist regulations. It was recorded that in the State of Qi, the system of capacity units could be divided into two categories. First, the authoritative category (公量), which was established by the royal court. In this system, the units consisted of dou (豆), ou (区), fu (釜), and zhong (钟) where1 ou = 4 dou,1 fu = 4 ou, and 1 zhong = 10 fu. Secondly, the domestic category (家量) established by those gentry and senior officials co-existed with the authoritative category. In this system, the names of the units were the same as those in the authoritative category, where 1 ou = 5 dou, 1 fu = 5 ou, and 1 zhong = 10 fu. In 1857 three pieces of measuring apparatus in bronze were unearthed in China. Inscriptions on these apparatus revealed that they were made in the State of Qi, and were serving as standard measures for calibrating other measuring apparatus and grains in or out granaries. On one of these apparatus the institution of calibration and regulations of punishment were inscribed. According to the inscriptions, two of these apparatus named fu (釜) were measured around 20,000 ml, and the third one was huo(铆),

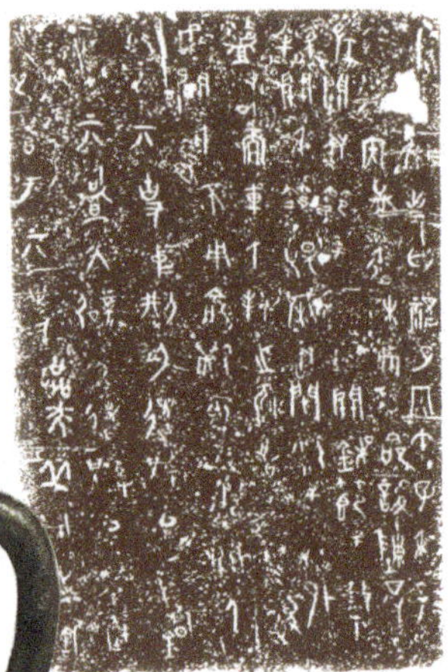

战国时期·齐　子禾子铜釜　器壁铭文阐述了三件量器的作用以及对违反规定、营私舞弊者的量刑处置。容 20460 毫升

A Zi He Zi Bronze Cauldron, 20,460 ml in capacity, in Qi of the Warring States Period, with inscriptions stating the functions of the three instruments and rules of penalizing those who violated the regulations in malpractices

战国时期·齐　左关铜铆　容 2070 毫升，约是釜容积的十分之一，相当于一斗量

A Zuo Guan Bronze Cauldron in Qi of the Warring States Period, 2,070 ml in capacity, about 1 dou or one-tenth of the capacity of either of the previous cauldrons which was measured around 2,000 ml. Recently more measuring apparatus made in Qi were unearthed in China. One of these apparatus was pottery on which sheng (升) was inscribed, and actually measured 200 ml. The rest of them were in bronze, and divided into two categories: small apparatus measured 200 ml, and big ones measured 1,000 ml, that is, 5 times of the small ones. So we can extrapolate that each of the big bronze measuring apparatus was equal to 5 sheng and each of the small ones was equal to 1 sheng. Then we can further know that— 1 huo = 10 sheng, and 1 fu = 100 sheng.

Based on the cross-checking between the measuring of those unearthed objects and records in historical documents, findings were obvious that with the changes of regimes of the State of Qi, the weighing and measuring system was also changed and improved continuously, from the units of 4 in the authoritative category to those of 5 in the domestic category, and further to the decimal system.

In some museums in China, some of exquisite bronze counterpoises (铜权) made in the State of Qi were stored. By referencing to historic documents we know that among those weighing units used in Qi, there was a unit of yi (镒) where 24 yi were equal to 1 jin (斤). A weighing of counterpoises with the inscription of yi (镒) found that 1 yi was equal to 370 grams. Therefore it is extrapolated that 1 jin (斤) in Qi was around 246 grams.

图间插文 Descriptions matching the Pictures

同时出土于齐国故地的三件铜量器：子禾子釜、陈纯釜、左关鋓。器壁铭刻说明它们是置于与其他诸侯国通商关卡上的量器。

The three bronze weighing instruments of the State of Qi, unearthed at the same time: Zi He Zi Cauldron, Chen Chun Cauldron and Zuo Guan He. Characters inscribed on the instruments indicate that they were used for commercial purposes in trade with the other states.

战国时期·齐 陈纯釜 器壁铭刻7行，除阐述对量器的管理制度外，还将监制和冶器者人名铸于器上，即物勒工名，以明确责任，保证质量。容20580毫升

Chen Chun Cauldron, 20,580 ml in capacity, in Qi of the Warring States Period, with 7 lines of inscriptions stating the regulations of the implementation of the weighing instruments and the names of those who were in charge of the production and inspection, for the purposes of clarifying responsibilities and assuring qualities

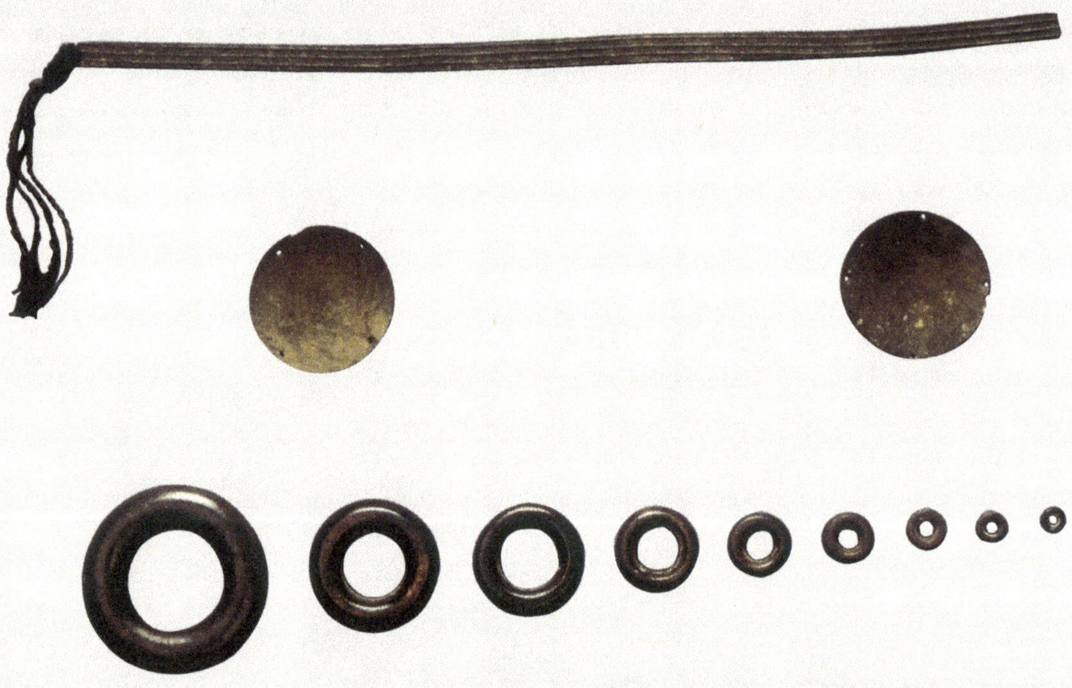

战国时期・楚 竹衡杆・铜环权 铜环权最大重125克，最小重0.6克，总重250克，合1斤
A bamboo beam and bronze weights in Chu of the Warring States Period, weighing from 0.6g to 125 g, totaling 250 g (about 1 jin)

（2）楚国的度量衡

楚国地处南方，位于长江中游江汉平原的沃土上。楚国经济繁荣，文化以及工商业都很发达，以青铜冶炼最为著名。楚国的都城"郢"是当时著名的商业城市。

近年来楚国的金属货币大量出土，除有铜铸币之外，还有金币，称"上币"。黄金价值高，交易时要经过称重后才进入流通，故学术界又称它为"称量货币"。为适应需要，小型天平便成为普遍使用的称重器具了。在这类小型天平上使用的环形砝码，据不完全统计先后已出土了400余件。目前所见最完整的一套包括衡杆、2个铜盘和10枚环形砝码。砝码的重量，大体以倍数递增，分别为一铢、二铢、三铢、六铢、十二铢，一两、二两、四两、八两和一斤。一铢约重0.69克，一两约15.5克，一斤约250克。十枚相加约500克，是楚制的二斤。

当时不仅广泛使用衡器，而且有了初期的杠杆理论。战国时期的科学巨著《墨经》中有一段关于杠杆力距平衡原理的论述："衡而必正，说在得衡。加重于其一旁必捶，权重相若也。相衡，则本短标长。两加焉重相若，则标必下，标得权也。"中国历史博物馆藏有一支战国时期的铜衡杆，正中有拱肩提纽和穿线孔。一面刻贯通上下的十等分线，全长相当于战国时的一尺（23.1厘米），每等分当为一寸。形式既不同于天平衡杆，又有异于秤杆，很可能是界于天平与杆秤之间的一种衡器。用这种衡杆称物，可以把被称物与权放在提纽两边不同位置的刻线上，即把衡杆的某一臂加长，这样，用同一个砝码就可以称出大于它一倍或几倍的重量。经过逐步发展，缩短重臂，加长力臂，就形成了秦、汉以后不等臂的杆秤了。

(2) The Measuring and Weighing System in the State of Chu (楚国)

The State of Chu was located on the fertile land of the Jianghan Plain, in the south of China. Chu was also a state with flourishing handicraft industry and commerce, and especially famous for its bronze industry. The excavated Tonglvshan (铜绿山) copper mine is the oldest and biggest one in the Chinese history. The capital of Chu, Ying (郢) was a famous commercial city at that time.

Recently large quantities of coins were unearthed in the former Chu, among which there were copper coins and golden coins called shangbi (上币 first-class coins) as well. Gold was of high value, so when it was used in trade, weighing was necessary. That is why gold is called "money of weighing" in academics. In the State of Chu, in order to use gold in trade conveniently, small balances were quite popular. According to incomplete figures from what has been unearthed up to now, there are more than 400 pieces of ring-shaped weights used on these small balances. The most intact balance unearthed comprised a piece of arm, two copper pans and 10 pieces of ring-like weights. Weighting units were inscribed on each of these weights as follows — 1 zhu (铢), 2 zhu, 3 zhu, 6 zhu, and 12 zhu; 1 liang (两), 2 liang, 4 liang, 8 liang and 1 jin (斤) .The weight marked 1 zhu weighed 0.69 g, the one marked 1 liang weighed 15.5g, and the one marked 1 jin weighed 250 g. So the whole set of the weights weighing around 500g, which is equal to around 2 jin, the scale prevailing in the State of Chu.

During the Warring States Period, not only the measuring apparatus were popular, but also theories for balancing were being developed, with the principle of balance. The renowned scientific works of that time — *Mo Jing* (《墨经》) recorded the common understanding of the principle of balance. In the Chinese History Museum there is a bronze beam (铜衡杆) made in the Warring States Period, which displays Chinese people's understanding of the principle of balance. At the middle of the beam a pivot is set and a hole drilled. On one side of the beam 10 graduations were engraved. The length of the beam measured 23.1cm, with each scale equal to 1 chi (尺) of that time. The beam reveals that this measuring instrument is neither an equal-arm balance, nor a cheng (秤杆). A hypothesis is that when applied this instrument in weighing, the object and the weight could be hung on each side of the beam, then by observing the positions of the object and the weight, the weight of the object could be calculated. When the length of the arm is increased, this instrument would become a cheng, which became popular in the periods after the Qin and Hand dynasties.

战国时期 • 楚 两组铜环权

Two sets of bronze weights in Chu of the Warring States Period

战国时期·楚 铸造环权的铜范
A bronze mould for casting weights, in Chu of the Warring States Period

古埃及 天平衡杆
Balance beams in ancient Egypt

战国时期·楚 "王"铜衡杆(两支)分别刻 10 等分和 20 等分线,长 23.1 厘米。可作不等臂天平衡杆使用
Two "Wang"(Lord) bronze beams in Chu of the Warring States Period with scales of 10 and 20, and a length of 23.1 cm respectively, for parallel weighing purposes

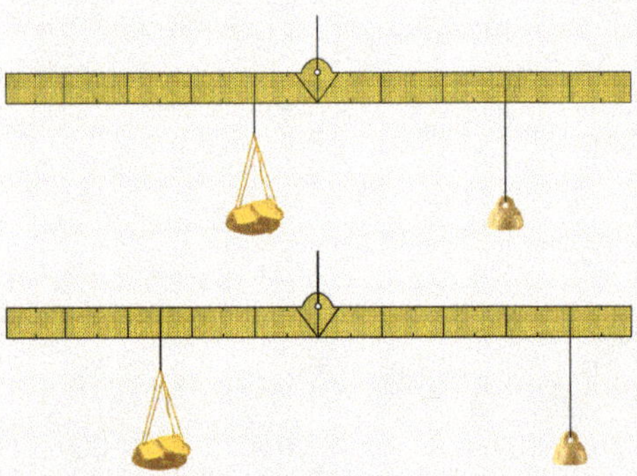

"王"铜衡杆使用示意图
An illustration of using the "Wang" (Lord) bronze beams

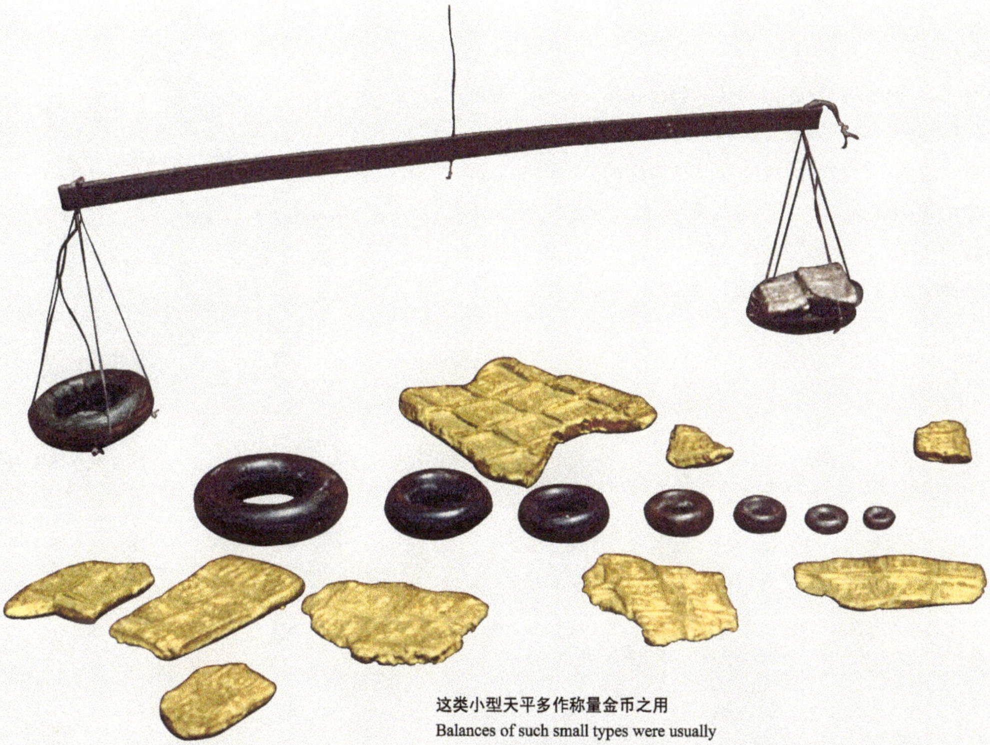

这类小型天平多作称量金币之用
Balances of such small types were usually used for weighing gold coins

战国时期・楚铜量 分别容 1110 毫升(5 升)、200 毫升(1 升)
Bronze weighing instruments in Chu of the Warring States Period with capacities of 1,110 ml (5 sheng) and 200 ml (1 sheng) respectively

古埃及 量器
An instrument for capacity in ancient Egypt

(3)赵国的度量衡

赵国位于北方。我们所见赵国的器物有两件铜权，一件上面刻有年代，监造者的官职、姓名以及校准的方法，并自铭为“禾石”，是一件用来称量谷物、自重120斤的大型铜权。权实重约30350克，按120斤折算，1斤约为253克。另一件自铭为“半石”(器已流失)。此外，赵国故地还出土了一大批刻有自重铭文的金银饰品，其单位有“两”“铢”。通过对这些金银饰件的测量，1两折合约15.6克，1斤约250克。

战国时期·赵“公朱半石”铜权 称禾谷的专用铜权，自铭“半石”，即60斤

“Gong Chu Ban Dan” bronze weight in Zhao of the Warring States Period specially for weighing grain, with inscriptions of “Ban Dan” (60 jin)

战国时期·赵 禾石铜权 铭文大意是：此为官方颁发，专用作称禾谷的120斤权。重30350克

He Dan Bronze Weight of 30,350 g (120 jin) in Zhao of the Warring States Period with inscriptions as officially decreed objects specially for weighing grain

战国时期·邹 陶量 底部有戳式印文“廪”字。当为国库仓廪所用的量。器容20000毫升

A pottery weighing instrument of 20,000 ml, in Zou of the Warring States Period with a stamp of “Lin” at the bottom, which was supposedly used in the state storehouses

图间插文 Descriptions matching the Pictures

春秋战国时期许多器物上刻有容量、重量的铭文。为研究此期间各诸侯国度量衡单位制提供了实证依据。

During the Spring and Autumn and Warring States Periods, there were many objects on which inscriptions were carved to indicate the capacities and weights. They serve as sizeable evidence in researches on the measuring and weighing systems in the states.

(3) The Measuring and Weighing System in the State of Zhao (赵国)

The State of Zhao was located in the north of China. Among those unearthed measuring-related objects made in Zhao, there were two pieces of bronze weights. One of them named Hedan (禾石), 30,350 g in weight, was engraved with the casting date, the official position and name of the man who was supervising the casting of the weight, and the method used for the adjusting of the weight. The name of the weight tells us that it was for the weighing of grain. The weight of the bronze, 120 jin was also engraved on it. So the weight jin prevailing in the State of Zhao was equal to 253 g. The other bronze weight was named Bandan (半石). Besides, a lot of ornaments were unearthed in the former Zhao, on which the weights were marked. The weighing units on these ornaments include liang (两) and zhu (铢). Comparing the weights marked on these ornaments with the actual weights in metric, we are able to calculate that 1 liang of Zhao was equal to 15.6 g, and 1 jin of Zhao was equal to 250 g.

战国时期 • 韩 陶量 器口沿有"廪""阳城"等铭刻，分别容 1670 毫升、1690 毫升
Pottery weighing instruments of 1,670 ml and 1,690 ml respectively, in Han of the Warring States Period with stamps of "Lin" and "Yang Cheng" at the mouths

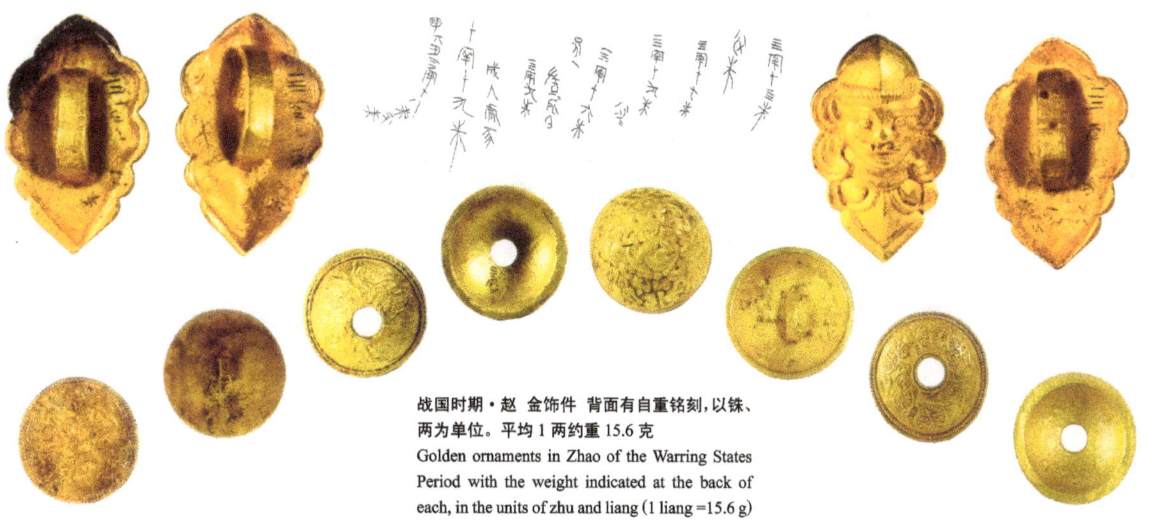

战国时期 • 赵 金饰件 背面有自重铭刻，以铢、两为单位。平均 1 两约重 15.6 克
Golden ornaments in Zhao of the Warring States Period with the weight indicated at the back of each, in the units of zhu and liang (1 liang =15.6 g)

（4）诸侯割据，度量衡制度混乱

春秋战国时期历时500多年，大大小小的诸侯国多达数十个，它们形成了一个个独立的中央集权国家。经过不断的兼并战争，到了战国后期形成了七雄争霸的局面。在此遴选三四个战国晚期典型的诸侯国为例，说明各国制度既有不同，又逐步趋于统一的特点，而另一方面又说明了这时期制度的复杂和混乱。

各国计量俸禄的单位不统一：卫国用“盆”，楚国用“担”，魏、齐等国用“钟”，秦国用“石”。从留存至今的相关实物来分析，各国度量衡单位名称、器物形制、单位量值、管理制度等方面也都存在着较大的差异。容量单位除战国后期各国比较普遍使用的升、斗之外，魏国用“镒”，燕国用“孛”“鹄”。即使单位名称相同，各国单位量值也存在差异。以1升单位为例，齐国约合200毫升，楚国约合226毫升，赵国约合175毫升。

然而，到了战国后期，由于商业和交通的发展，各诸侯国之间的联系大大加强，各地区、各部族在经济上的交往和相互依赖也更加密切。由于经济的发展，新兴地主阶级的改革和封建兼并战争的不断发生，更由于文化的交流、融合，使得原来封闭式的割据状态受到了很大冲击。促使各国度量衡发展逐步走向统一，这也为秦始皇统一度量衡奠定了基础。

战国时期·燕 铜壶 器口刻此壶容一孛五鹄
A bronze pot in Yan of the Warring States Period, with inscriptions of the capacity of 1 gu 5 ju carved at the mouth

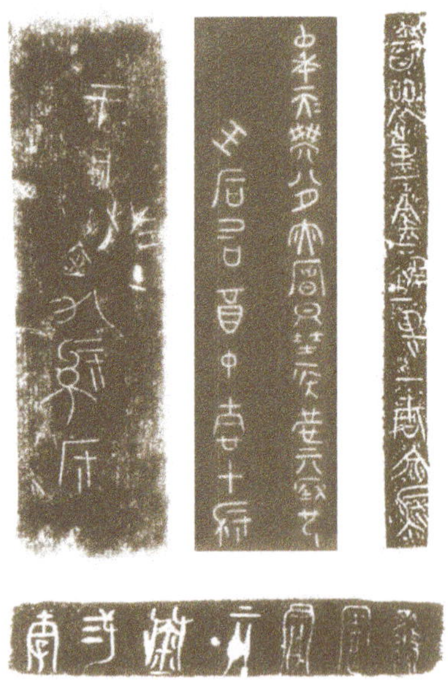

战国时期·燕铭文拓本 所见多件铜器上有以“孛”为单位记容积。经综合分析，1孛约容1766毫升
In Yan of the Warring States Period, inscriptions of units of capacities on each. Synthetical study shows that the unit of weight was about 1,766 ml

(4) The Confusions in the Measuring and Weighing Systems

The Spring and Autumn and Warring States Periods had lasted for more than 500 years, during which more than 10 states established centralized state powers successively. After many years of furious warfare aiming at expansion and annexation, seven out of the 10 states survived by the Warring States Period. The situations in some of the States were cited below to illustrate the features of the measuring and weighing system of that time.

Units of officials' salaries were different among different states. In the State of Wei, the unit was pen (盆); in the State of Chu, it was dan (担); in the States of Wei and Qi and some other states, it was zhong (钟); and in the State of Qin, it was dan(石). An analysis based on unearthed measuring apparatus of that time reveals that besides names of units, shapes and standards of apparatus, values of units, and management systems were also different among different states. As for the units of capacity, sheng (升) and dou (斗) were popular in these states while there were yi (镒) in the State of Wei, gu(孛)and ju (鹌) in the State of Yan. Even if some units were commonly used among different states by the name, values of these units might be different. Take sheng, for example, it was

战国时期・中山 铜壶 圈足铭刻记壶的重量，单位用"石""刀"
A bronze pot in Zhongshan of the Warring States Period, with inscriptions of the weights on the feet, in the units of "dan" and "dao"

equal to 200 ml in Qi, 226 ml in Chu, and 175 ml in Zhao.

Towards the end of the Warring States Period, owing to the progress of commerce and transportation, ties among the states were strengthened. At the same time the economic development, endless wars, communication and integration among diverse cultures were interwoven in the situations where each separated regime was isolated from the others. Generally speaking, the momentum of tendency to unify the measuring and weighing systems among different states was strong, which laid the foundation for the standardization of the system of measures and weights in China.

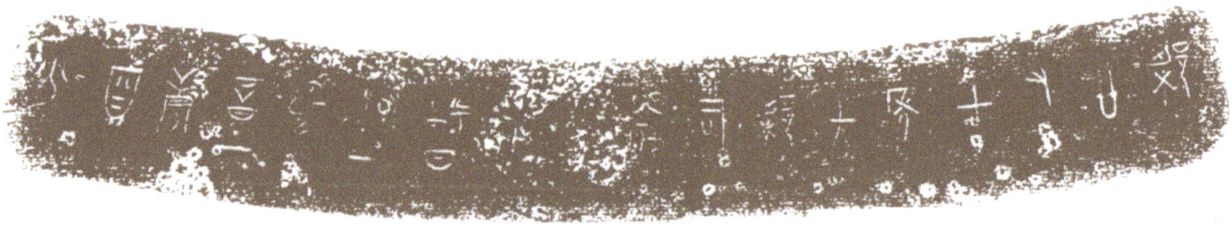

战国时期·魏 铜鼎 自铭容"四分"即四分之一禽(鬶)。实容 1800 毫升，正是一鬶的四分之一
A bronze tripod (Ding) in Wei of the Warring States Period with inscriptions of the capacity which was actually 1,800 ml, just a quarter of ji

A. 铜钫
Bronze pots

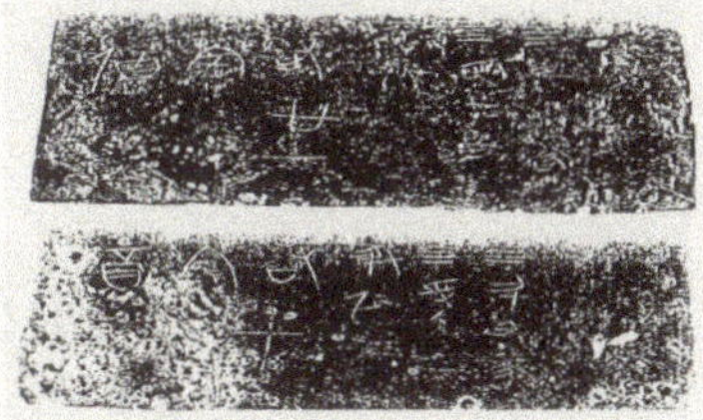

B. 铭文拓本
Inscriptions in the form of rubbings

东周时期 铜钫 所见数件铜钫皆刻铭容××斗、重××寽，折算1斗约容7700毫升，1寽约重1224克
Bronze pots (Fang) of the Eastern Zhou, all carved with capacities in units of dou, and with weights of lue. The capacity of 1 dou was about 7,700 ml, and 1 lue about 1,224 g

战国时期 彩绘骨尺 长22.5厘米
A bone ruler, 22.5 cm in length, with colored patterns in the Warring States Period

东周时期 铜尺 长23.1厘米
A bronze ruler, 23.1 cm length, in the Warring States Period

战国时期·魏 铜鼎 自铭容半齋实容3600毫升，1齋7200毫升
A bronze tripod (Ding) in Wei of the Warring States Period with inscriptions of the capacity of 3,600 ml, half of ji . 1 ji was about 7,200 ml

THE HISTORY

OF ANCIENT CHINESE

EASURES AND WEIGHTS

5

THE HISTORY
OF ANCIENT CHINESE
MEASURES AND WEIGHTS

中国古代计量史

五、从商鞅变法到秦始皇统一度量衡

Chapter Five

From Shang Yang Reform to Qin Shi Huang's Standardization of the Measuring and Weighing Systems

春秋战国之际，各诸侯国顺应时代的发展先后进行改革和变法，由分裂割据状态下的奴隶社会陆续向封建社会过渡。秦国地处西北边陲，在从奴隶社会向封建社会转变过程中，落后于东方各国。公元前384年，秦献公即位，采取了一些改革措施，如开辟交易市场，注重发展经济等，国势有所好转。经过几代人的努力，秦国终于从弱到强，为秦始皇统一全国打下了基础。

During the Spring and Autumn and Warring States Periods, in order to cope with the new situation, each of the States initiated reforms, at one time or another. Through these reforms the feudalist social system gradually came into being. The State of Qin, located at the northwest of China, was falling behind the other states in the evolution from slave society to feudal society. In the year 384 B.C., Lord Xian (献公) ascended the throne and began to carry out reforms. Measures such as opening up markets and encouraging commercial activities were taken, and the whole situation of the state started to take a favorable turn. Through continuous efforts of generations, the State of Qin became rich and strong, which laid the foundation of unifying the Warring States in the years to come.

商鞅辅助秦孝公变法，提出统一度量衡
Shang Yang, while helping Lord Xiao with the Reformation, advocated to unify the measuring and weighing systems

1. 商鞅变法与商鞅铜方升

继秦献公之后，改革取得明显成效的是秦孝公（前361～前338在位）。孝公广招天下贤士，决心变法图强。商鞅在这种政治条件下来到了秦国。商鞅到秦国不久就得到孝公的重用，经过五年时间，变法取得初步成功。公元前352年商鞅被提升为大良造，官职相当于中原各国的相国兼将军。公元前350年又进行了第二次变法，其中与度量衡相关的有以下两点：①废井田，开阡陌封疆。对田亩制做了新的规定，将过去的100方步为亩改为240方步为亩。定出丈量土地的办法和标准，即“举足为跬，倍跬为步”。一步为六尺。从而扩大了耕地面积，有利于农业的发展。在中央直接管辖下，田亩大小有了统一的标准，为平准赋税提供了有利条件。商鞅之法“步过六尺者罚”。当时新的田亩制已成为不可违抗的法律条文。②统一度量衡，颁布度量衡标准器。公元前348年，秦国进一步推行赋税制度，需要有统一的度量衡作保证的有以下几方面。秦法规定，允许地主以个人名义占有耕地和宅地，但要严格按所占土地面积缴纳赋税。商鞅主张“禄厚而税多”，根据收入按比例上交国税。秦国设有“治粟内史”，主要任务是征收以粮食为主的土地税，用于国库开支及官吏的俸禄。又设“少府”，主要征收人口税、手工业税和商业税。此外，地方官吏每年还要向中央政府“上计”（相当于今天的统计）。过去士大夫阶层相互倾轧，肆意扩张各自的势力，常常在各自的庄园内另设“家量”，与“公量”相抗衡。度量衡的不统一，给变法革新造成许多障碍和困难。商鞅当机立断，颁布了统一度量衡的命令，随后又亲自设计并监督制造了一批标准器发到全国各地。现由上海博物馆收藏的“商鞅铜方升”，便是珍贵的实物资料。此器三面及底部均刻有铭文，左侧为：“十八年（前344），齐遣卿大夫众来聘。冬十二月乙酉，大良造鞅。爰积十六尊（寸）五分尊（寸）一为升。”与柄相对的一面刻“重泉”，是战国时的地名，即方升使用地。底部是公

战国时期·秦 商鞅铜方升 商鞅于公元前344年督造的标准量器。铭文记：“十六尊（寸）五分尊（寸）一（16.2立方寸）为升。”经实测后验算，方升深1寸，宽3寸，长5.4寸，得1尺约合23.1厘米，1升约合200毫升

Shang Yang Bronze Fang Sheng in Qin of the Warring States Period, produced in 344 B.C. under the supervision by Shang Yang. The inscriptions read, “16.2 cubic cun as 1 sheng”. Checking computations after an actual measurement showed that the fang sheng was 1 cun in depth, 3 cun in width and 54 cun in length. 1 sheng was equal to 200 ml

元前 221 年秦始皇统一全国后加刻的统一度量衡诏书。右侧为“临”字，系第二次使用地名。《史记》中有：孝公“十年，卫鞅为大良造。”方升正是商鞅任大良造八年之后督造的。统一度量衡在当时属重大事件，故在器物上刻铭记录。左侧铭文大意是：秦孝公十八年（前 344），齐国派遣由卿大夫等人组成的外交使团到秦国商讨包括有关度量衡等事项。方升自铭容积为 16 又 1/5 寸，即 16.2 立方寸。经测量研究得知当时秦国 1 尺约合 23.1 厘米，1 升约 200 毫升。

除商鞅铜方升之外，秦在战国时期制造的度量衡器主要还有铜质权。铜权周围有不同时期铸刻的铭文，铭文的内容包括：监造、铸造者官职、姓名、铸造年代、用途、使用地和权的自重等。如“高奴，禾石。”“高奴”是地名，“禾石”即称粮食的 120 斤权。从所刻铭文可知，秦在战国时期制造度量衡器具时，已做到管理有序、制度严明。秦始皇统一全国度量衡后，将战国时秦国旧有的度量衡器都做了重新检测，并均加刻上统一度量衡诏书。秦二世继位后，检测各地的度量衡器并再次颁发诏书，亦刻于一侧。此权自始铸至秦二世元年，三次铸刻铭文，在数百年中长期作为标准器使用，反映了从战国时期秦国至秦王朝，一直保持着统一的度量衡制度。

近年还出土了刻有自重的铜器和金银饰件。推算出当时 1 斤约合 253 克。

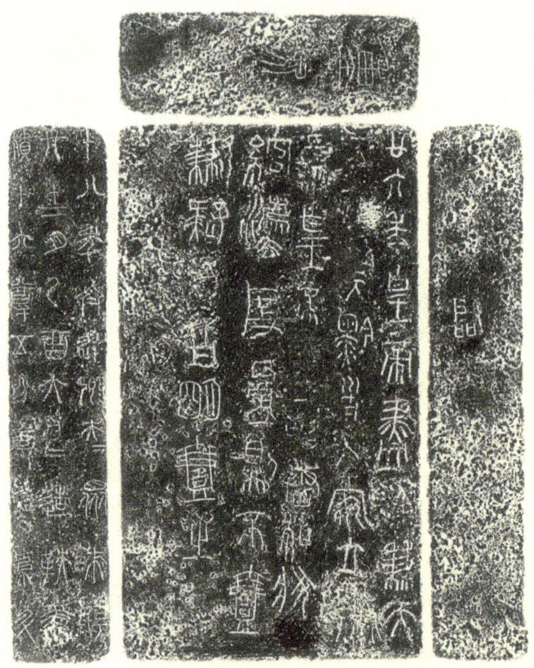

商鞅铜方升铭文拓本　左侧铭文大意是：（秦孝公）十八年（前 344），齐国派卿大夫来秦国。大良造（商）鞅督造方升。容积 16.2 立方寸。底部的铭文是 100 多年后，秦始皇统一全国，又加刻的统一度量衡诏书

Shang Yang Bronze Fang Sheng with inscriptions in the form of rubbings: on the left, they read, in the 18th year of Lord Xiao (344 B.C.), the State of Qi dispatched a Qing Da Fu to Qin. Da Liang Zao Shang Yang produced fangsheng. The capacity was 16.2 cubic cun. The inscriptions at the bottom were contents of the imperial edict of additional inscriptions on the unified system of measuring and weighing, after Qin Shi Huang unified China about 100 years later

这类小型量器多用于分配士卒每天的口粮

This type of capacity instrument was normally used for allotting daily rations to soldiers

战国时期·秦 禾石铜权 制器时刻"高奴"(地名)"禾石"等。秦始皇时，在一侧加刻统一度量衡诏书。秦二世继位，再次加刻二世诏书。重 30750 克

A He Dan bronze weight in Qin of the Warring States Period, 30,750 g in weight. Carved with "Gao Nu" (the name of a place) and "He Dan" when produced, the weight had the imperial edict of the unified measuring and weighing system added on one side during the Qin Shi Huang reign, and an additional imperial edict was added when Qin Ershi (the heir) succeeded to the throne

1. Shang Yang's Reformation and Shang Yang Bronze Fang Sheng

After Lord Xian's (秦献公) reform, Lord Xiao (秦孝公, reigned 361B.C.-338 B.C.) also initiated reforms and achieved remarkable successes. He invited and gathered scholars from all over to help him, and Shang Yang (商鞅) was the preeminent one among them. Shortly after his arrival, Shang Yang was granted a crucial position. Five years later, in 352 B.C. he was promoted to the position of Da Liang Zao (大良造). This position was similar to the prime minister and concurrently Commander-in-Chief in those states in the central plains of China. In 350 B.C., the second reform was carried out, in which the measuring and weighing system was involved. Two relevant aspects were—Firstly, abolishing the existing well-field system and legally recognizing private ownership of land. The unit in measuring the area of land, namely, mu (亩) was redefined as 240 square bu (方步), compared with 100 square bu in the past. Meanwhile the method and standard for the land measurement were specified: "One step was counted as kui (跬), and double kui was defined as bu". One bu was equal to 6 chi (尺). Under the central control, such a reform was favorable for the development of agriculture and provided a solid base for the standardization of levying tax. Secondly, abolishing the domestic measuring system to unify the standards of measure of fields, and adopting standard measuring apparatus. Shang Yang supervised the establishment of items in the Law of Qin that encouraged ownership of land and taxation accordingly, and "high salary and high income taxes". Because of local lords' competition, using domestic measuring system, the authoritative measuring system (公量) was challenged. This was not allowed in the central government. According to *Shi*

Ji (《史记》), to deal with these issues, Shang Yang designed a series of standard measuring apparatus and supervised the production of them and delivery throughout the state, among which there was a bronze square sheng measure called the Shang Yang Bronze Fang Sheng (商鞅铜方升). This object is now stored in the Shanghai Museum. On the left side the inscription explains that this object was cast under the supervision of Shang Yang in 344 B.C., and the capacity was 16.2 cubic cun. On the other side a place name called Zhong Quan (重泉) was engraved, which means that this measure was first assigned to use in that place. On the right there engraved another place name, Lin (临), where the measure was secondly assigned to use. On the bottom the imperial edict of Qin Shi Huang ordering the standardization of measuring and weighing system was engraved in 221 B.C. after Qin Shi Huang unified the Warring States as China. The actual capacity of this object fang sheng (方升) was 16.2 cubic cun (寸). Based on specific measuring, it is found that the value of 1 chi in the Qin and Han dynasties was about 23.1 cm, and 1 sheng was about 200 ml. Another important measuring instrument made in the State of Qin in the Warring States Period was a bronze counterpoise. As usual there are inscriptions around the counterpoise engraved in different dates. The first inscription includes such contents as the name and title of the supervisor, the casting time, the usage and the place where the counterpoise was used, and the weight of the counterpoise itself. For example, Gaonu (高奴), hedan (禾石), where

A. 铸造时所刻铭文
Inscriptions carved when the instrument was made

B. 器口沿加刻容积拓本
Rubbings of the inscriptions of the capacity on the rim

战国时期 秦公簋 秦景公即位时(前 576)铸造。至秦始皇时又以统一的度量衡制校测,在器和盖的口沿上刻"一斗七 升大半升"。器容 3560 毫升,1 升合 201.5 毫升

Qin Gong Gui in the Warring States Period made in the year Qin Jing Gong ascended the throne (576 B.C.) The capacity (3,560 ml) was checked in the times of Qin Shi Huang , with the unified system of measuring and weighing which was carved on the instrument and its lid as "1 dou, 7 sheng and greater than half of 1 sheng". 1 sheng was equal to 201.5 ml)

Gaonu was the name of a place; he clarified the usage of the counterpoise, that is, for the weighing of grain. Dan is a unit of weight equal to120 jin at that time. From the inscription we know that in Qin times the making of measuring instruments was strictly regulated by the government. After Qin Shi Huang succeeded in the unification of China, the government of Qin checked and calibrated old measuring instruments made in the Warring States Period, and engraved the imperial edict of Qin Shi Huang ordering the standardization of measuring and weighing system on these instruments. Besides, on this counterpoise the imperial edict of "the Second Emperor" was also engraved after he decreed to check measuring instruments of the country. Inscriptions on this counterpoise reveal that within hundreds of years from the time when it was cast to the first year of the First Emperor (Qin Shi Huang), this object had been accepted by authorities to be a standard apparatus. It is then proven that the measuring and weighing system ever since the Qin times had been a unified one throughout this long period of time. We can also calculate the value of jin from some recently unearthed silver and gold ornaments made in the Qin times, and the result is that 1 jin in the Qin times was equal to 253 g or so.

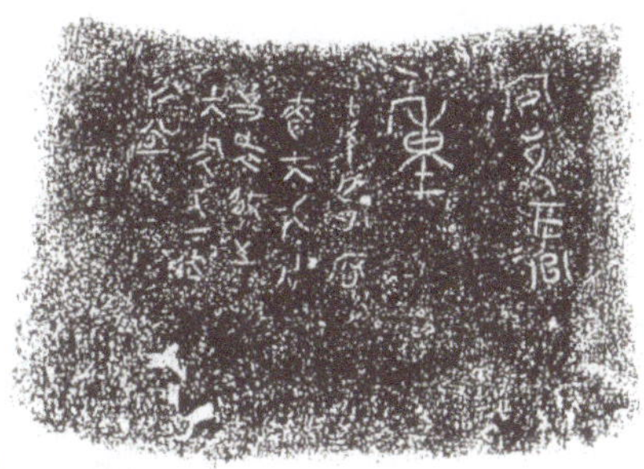

A. 铸造时所刻铭文
Inscriptions carved when the instrument was made

B. 器口沿加刻容积拓本
Rubbings of the inscriptions of the capacity on the rim

战国时期 • 魏 安邑下官铜钟 原有魏国时所刻器容积的铭文。秦获此器后，按秦制在口沿上加刻了"十三斗一升"。实容 26400 毫升
An Yi Xia Guan Bronze Cup in Wei of the Warring States Period with original inscriptions of the capacity in the State of Wei. An additional inscription of "13 dou and 1 sheng" was carved on the rim after it was

2. 秦始皇统一度量衡

公元前 221 年秦始皇统一全国，结束了长达 548 年诸侯割据、战乱频繁的动荡局面。为了巩固国家政权，秦始皇采取了一系列重大措施来加强专政，如在全国范围内推行以郡县制为基础的中央集权制，皇帝为最高统治者等。为了保证各种赋税的征收，维护封建秩序，下令在原有秦国法律基础上制定了一套更严密、更完善的法律条文。如《田律》规定按受田之数缴纳赋税。《仓律》规定，出入仓廪的粮食必须严格计量。《金布律》对布匹的长、宽有严格尺寸规定。《工律》对官营手工业生产各种产品要求规范化、标准化，以及对度量衡器具的使用、检定、管理都有严格规定等。以上各项条款都必须建立在统一度量衡基础之上。在秦始皇统一全国之前，度量衡的状况是，尺、寸、升、斗、斤、两等常用单位已在多数诸侯国中通行，单位量值差别也不太大。但由于各国发展不平衡，地区不同，改革力度不同，仍有一些国家和地域保留着旧的单位制，严重地影响着统一赋税、俸禄和奖惩制度等的贯彻执行。在秦始皇统一全国后，统一度量衡便成为刻不容缓的大事。

公元前 221 年，秦始皇统一全国后，颁诏书统一度量衡
Qin Shi Huang issued his imperial edict of the unified measuring and weighing systems after he unified China in 221 B.C.

2.The Standardization of the Measuring and Weighing Systems

In 221 B.C., after successive annexations and terminations, Ying Zheng (嬴政), the King of Qin, successfully unified the seven Warring States and the Empire was China. Ying Zheng established a highly centralized feudal regime, after a chaotic period of 548 years, which was the first of its kind in the Chinese history. Ying Zheng addressed himself the First Emperor (始皇 Shi Huang). In order to strengthen the regime, a lot of measures were taken, including a system of prefectures and counties (郡县制). In such centralized governmental structure, the Emperor was the paramount leader. The First Emperor ordered to issue decrees to recognize private ownership of land in the empire. Every owner of land should report the area of land possessed to the local government, which was the base for the amount of taxes to be levied. The First Emperor ordered to make a set of laws on the basis of the legal systems of the Qin regime to maintain the feudal order in every aspect of social life. Among these laws was *the Law of Land*(《田律》),which stipulated that owners of field must pay tax according to the area of land they owned. There was *the Law of Granaries* (《仓律》), which stipulated that grains in or out the granary must be preciously weighed. There was *the Law of Cloth* (《金布律》), which stipulated the length and width of cloth sold on markets. There was *the Law of Handicraft* (《工律》), which stipulated that handicraft of all kinds must be made according to certain standards. In all of these laws the measuring and weighing system was involved, including stipulations on application, inspection, and management of it.

Before the unification of China, units such as chi (尺), cun (寸), sheng (升), dou (斗), jin (斤), liang (两), etc. were used in most of the states, and there was no noticeable difference between the value of each unit in these states. But in order to make these laws to be enforced effectively, legal standardization of measures and weights was a must.

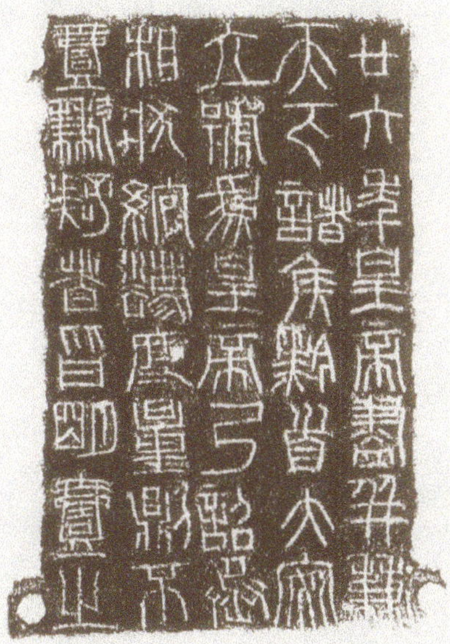

秦始皇统一度量衡铜诏版及拓本

A bronze plate and rubbings of the unified measuring and weighing system after Qin Shi Huang unified China

3. 秦始皇统一度量衡的具体措施

(1)颁发统一度量衡的法令

公元前221年(秦王政二十六年)秦王嬴政统一了中国，首立皇帝称号。凡他所规定的条例为“制”，下达的命令为“诏”。统一度量衡就是以诏书的形式下达的，也是目前留存下来的第一道“诏书”。诏书全文如下：“廿六年，皇帝尽并兼天下诸侯，黔首大安，立号为皇帝，乃诏丞相状、绾，法度量则不壹歉疑者，皆明壹之。”大意是：他继位的第26年，兼并了各地诸侯国。百姓得以安居乐业，自立皇帝称号。下诏书给两位大臣，务必把全国的度量衡都统一起来。诏书全文共40个字，简要地说明了统一度量衡的历史背景和对统一的要求。铜铁质器上的诏书是直接刻、铸在器物上的。陶质量器是预先烧制成10个印戳分别钤印在泥坯上，连成一篇诏书后再烧制成器。此外还出土了数件铜诏版，四角有孔，可能是钉在方形木制大型量器上的，木质易腐，故仅留下诏版。由皇帝颁发的统一度量衡命令，有力地保证了这项措施顺利推行和坚决贯彻。

3. Actual Methods Taken for the Standardization of Measures and Weights

(1) Issuing Decrees to Standardize Measures and Weights

Regulations from Qin Shi Huang ,the Emperor were called zhi (制) and decrees from him were called zhao (诏, imperial edict). The ordering for standardization of measures and weights containing 40 Chinese characters, was the Emperor’s first imperial edict since he was crowned. The common people would live and work in peace and contentment, and he thus awarded himself a new title Huang Di (皇帝, Emperor). Now he ordered two of the ministers to standardize the measures and weights. This edict was on a piece of pottery slab, which was first sealed by 10 pieces of stamps with each engraved with a part of the edict and then baked. Beside this pottery edict, there were also several pieces of bronze edict unearthed. They were found previously nailed on wooden measuring apparatus that were rotten off, with these bronze slabs left. The edict ordering the standardization of measures and weights guaranteed the implementation of it.

秦代 铁石权 权身镶嵌铜诏版。重32500克

An iron dan weight, 32,500 g in weight, in the State of Qin, mounted with a bronze plate of the imperial edict

秦代 铜诏版 同时刻有秦始皇和秦二世统一度量衡诏书

A bronze plate of the imperial edict of the unified measuring and weighing system, in the State of Qin, issued by Qin Shi Huang and Qin Er Shi

秦代 铜权 权身刻秦始皇及秦二世诏书。重 7620 克，当秦时一钧(30 斤)权

A bronze weight in the State of Qin, 7, 620 g in weight, carved with imperial edict issued by both Qin Shi Huang and Qin Er Shi. Rubbings of the imperial edict stated that 1 jun was equal to 30 jin

秦代 秦二世时铸造的一斤铜权，同时刻秦始皇和秦二世诏书，重 240 克
A 1-jin bronze weight made during the Qin Er Shi times, 240 g in weight carved with imperial edict issued by both Qin Shi Huang and Qin Er Shi

秦代 秦始皇和秦二世诏一斤铜权，重 250 克
A 1-jin bronze weight, 250 g in weight carved with imperial edict issued by both Qin Shi Huang and Qin Er Shi

秦代 秦始皇和秦二世诏九斤铜权，重 2270 **克**
A 9-jin bronze weight, 2,270 g in weight carved with imperial edict issued by both Qin Shi Huang and Qin Er Shi

秦代 秦始皇和秦二世诏九斤铜权，重 2300 **克**
A 9-jin bronze weight, 2,300 g in weight carved with imperial edict issued by both Qin Shi Huang and Qin Er Shi

秦代 秦始皇诏八斤铜权，重 2064 克
An 8-jin bronze weight, 2,064 g in weight carved with imperial edict (rubbings) issued by Qin Shi Huang

秦代 秦始皇诏五斤铜权，重 1264.6 克
A 5-jin bronze weight, 1,264.6 g in weight carved with imperial edict (rubbings) issued by Qin Shi Huang

120 斤
120 jin

30 斤
30 jin

20 斤
20 jin

16 斤
16 jin

（2）沿用战国（秦）时度量衡制度、法规

秦始皇统一度量衡，沿袭了自商鞅以来制定，并在秦国已推行了100多年的法规、制度和标准。全国统一后，更把秦制强制推广至全国。这一点已从“商鞅铜方升”“高奴禾石铜权”的铭文上得到了证实。说明了秦对度量衡器的检测从未间断过，而且对统一量值的管理也是十分严格的。虽经历了几代人，100多年时间，度量衡三个量的单位量值都能一代一代传递下去，甚至被秦以后的汉、晋历代又延续了五六百年。

(2) Promoting the System of Measures and Weights of the Previous State of Qin

The standardization of measures and weights was actually to promote the system established in the previous State of Qin during the time of Shang Yang. Inscriptions on the Shang Yang Bronze Fang Sheng (商鞅铜方升) and Gaonu He Dan Bronze Weight (高奴禾石铜权) serve as proof of this purpose. Because the standard measuring apparatus had been inspected and calibrated consistently throughout the Qin times, and a set of laws and regulations had been strictly enforced, values of units could be passed onto following generations and continued to be used in the next dynasties of Han (汉代) and Jin(晋代) for another 500 to 600 years.

秦代 秦始皇诏一斤铜权
A 1-jin bronze weight carved with imperial edict issued by Qin Shi Huang

秦代 秦始皇诏一斤铜权
A 1-jin bronze weight carved with imperial edict issued by Qin Shi Huang

秦代 秦始皇诏二斤铜权
A 2-jin bronze weight carved with imperial edict issued by Qin Shi Huang

秦代 各种不同量值、不同形式的秦权
Various types of weights with different values in the Qin Dynasty

10斤
10 jin

10斤
10 jin

8斤
8 jin

1斤
1 jin

半两
1/2 liang

（3）制造和颁发度量衡标准器

流传至今的秦代度量衡器物共百余件，绝大多数刻有诏书，有的还加刻了秦二世诏书。秦二世诏特别强调统一度量衡是秦始皇的功绩，并下令必须把统一的制度继续贯彻下去。

秦权形制皆为半球形，顶部有鼻纽和穿孔，便于系绳。权的量值分石（120 斤）、钧（30 斤）、24 斤、16 斤、8 斤、5 斤、1 斤和半两等。据研究，秦权皆作砝码使用，即提纽在衡杆的中间部位，一端挂权，另一端挂被称物。衡平则称准。量器形制变化较多，除长方形的铜升外，小型铜量多呈椭圆形。

近年出土的秦代度量衡器，不仅数量多，分布范围也很广。除原秦国故地外，燕国故地、楚国故地、齐国故地、韩国故地、赵国故地都曾出土。有力地证明了秦代在辽阔的疆域内实现了度量衡的统一。

秦代 秦始皇诏铜量 容 485 毫升，即四分之一斗
A bronze instrument of 1/4 dou (485 ml) with imperial edict issued by Qin Shi Huang

秦代 秦始皇诏一升铜量 容 210 毫升
A 1-sheng bronze instrument, 210 ml in capacity, with imperial edict issued by Qin Shi Huang

秦代 秦始皇和秦二世诏铜量 容 980 毫升，即半斗
A bronze instrument of 1/2 dou (980 ml) with imperial edict issued by both Qin Shi Huang and Qin Er Shi

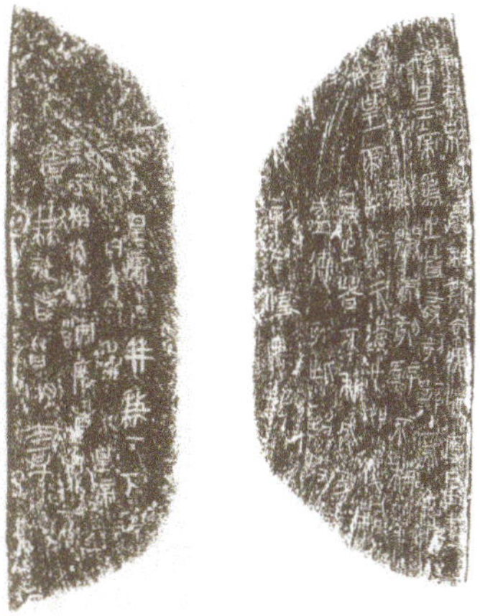

秦代 秦始皇及秦二世诏铜量 容 650 毫升，当三分之一斗
A bronze instrument of 1/3 dou (650ml) with imperial edict issued by both Qin Shi Huang and Qin Er Shi

秦代 秦始皇诏铜量 容 490 毫升，当四分之一斗
A bronze instrument of 1/4 dou (490 ml) with imperial edict issued by Qin Shi Huang

(3) Producing Standard Measuring Apparatus

Large quantities of measuring apparatus were made during the regime of Qin Shi Huang for the standardization of measures and weights, among which hundreds were still remained. Most of these measuring apparatus were engraved with the edict of Qin Shi Huang, some even with the edict of Qin Er Shi (秦二世诏书) additionally. The edict of the Qin Er Shi emphasized that the standardization of measures and weights was a great contribution of Qin Shi Huang , and made it clear that the unified measuring and weighing system must be carried out.

A weight made in the Qin times was usually half-globe in shape, with a knob and a hole on the top of it. All of these weights still remained were marked as dan (石, 120 jin), jun (钧, 30 jin), 24 jin, 16 jin, 8 jin, 5 jin, 1 jin and half liang (半两) respectively. According to research findings, Qin weights were used as counterpoises: the pivot was set in the middle of the beam, and a weight and the object to be weighed were suspended from each end of the beam.

Unearthed measuring apparatus of the Qin times are not only large in quantity, but also wide in scattering areas. Apart from the State of Qin, the main unearthed spots include the former Yan (燕国) territory, the former Chu (楚国) territory, the former Qi (齐国) territory, the former Han (韩国) territory, and the former Zhao (赵国) territory. All these prove that the Qin Dynasty had achieved the standardization of measures and weights within its vast territory.

秦代 秦始皇诏铜方升 容 215 毫升
A bronze instrument, Fang Sheng, 215 ml in capacity, with imperial edict issued by Qin Shi Huang

秦代 秦始皇诏陶量 器壁刻始皇诏书(残缺,经修复)。测其容 32000 毫升,当秦之 1 斛 6 斗
A pottery instrument in the times of Qin Shi Huang, carved with imperial edict (fragmented and repaired). The capacity was measured as 32,000 ml, about 1 hu and 6 dou in the Qin Dynasty

秦代 秦始皇诏陶量 器壁有始皇诏书用木戳钤印后烧制而成。容 2000 毫升,当秦 1 斗
A pottery with instruments in the times of Qin Shi Huang, with an imperial edict stamped on. The capacity was measured as 2,000 ml, about 1 dou in the Qin Dynasty

秦代 秦始皇诏陶量 容 990 毫升，即秦半斗
A pottery with instruments in the times of Qin Shi Huang , with imperial edict. The capacity was 990 ml, about 1/2 dou in the Qin Dynasty

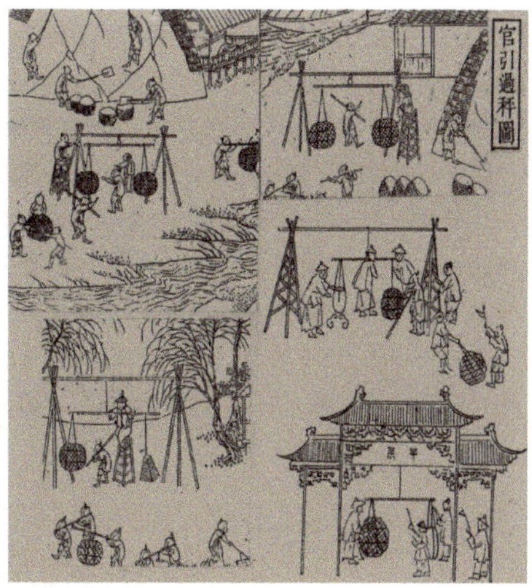

解盐称重图（清）盐一包重一引（200 斤）
A drawing of weighing salt in the Qing Dynasty(清代) . The weight of 1 pack was 1 yin (200 jin)

大型秦权使用示意图
A diagram in large scale of the application of the weights in the Qin Dynasty

15 世纪比利时市场上大型天平使用图
A diagram in large scale of the application of balance on a market in Belgium in the 15th century

1497 年德国纽伦堡天平浮雕
Rilievo of a German balance in 1497

(4) 实行严格的检定制度

为了保证"器械一量",秦始皇除了制造大量的有统一量值的度量衡器具发放到全国各地,制定了严格的检定制度外,还通过法律条文十分详细地规定了度量衡器具在使用中的允许误差范围。今综合整理如下。

衡器。石不准确,误差在 16 两以上,罚该官府官啬夫一副铠甲;不满 16 两而在 8 两以上者,罚一盾牌。半石不准确,误差在 8 两以上,罚一盾牌。钧不准确,误差 4 两以上,罚一盾牌。斤不准确,误差 3 铢以上,罚一盾牌。称黄金所用的小型权衡器不准确,误差在半铢(约 0.3 克)以上,罚一盾牌。

量器。桶(斛)不准确,误差在 2 升以上,罚一副铠甲;不满 2 升而在 1 升以上,罚一盾牌。斗不准确,误差半升以上,罚一甲;斗不满半升而在 1/3 升以上,罚一盾牌。半斗不准确,误差在 1/3 升以上,罚一盾牌。1/3 斗不准确,误差在 1/6 升以上,罚一盾牌。升不准确,误差在 1/20 升以上,罚一盾牌。

秦统一度量衡的各项措施,对其后 2000 多年封建社会度量衡制度产生了极其深远的影响。

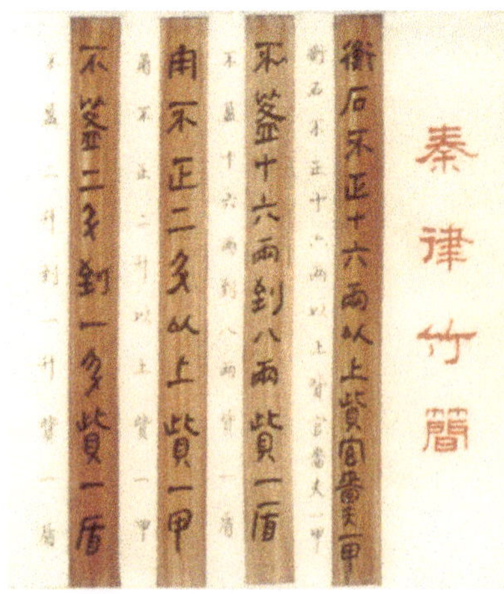

秦律规定,凡使用超出允许误差范围的度量衡器具者,按超差多少,分别处罚铠甲或盾牌
Legal regulations set in the Qin Dynasty that the penalty varied according to the seriousness of abusing the instruments of measures and weights: the overproof amount determined fines of an armor or a shield

(4) Implementing Strict Inspection and Calibrating Processes

To ensure the unity of apparatus, Qin Shi Huang ordered to produce in large quantities apparatus of measuring and weighing, and allocated them all over the country. Through verifications in the Law of Qin, it was regulated that mistakes were allowed to make within a range of limitations. The main points are summarized as follows.

Weighing Apparatus. For dan (石), if the mistake was more than 16 liang (两), the punishment for the official was the equivalent of a set of armor. If the mistake was less than 16 liang but more than 8 liang, the punishment was equivalent of a shield. For bandan(半石), if the mistake was more than 8 liang ,the penalty was a shield. For jun (钧),if the mistake is more than 4 liang ,the penalty was a shield. For jin(斤),if the mistake is more than 3 zhu (铢), the penalty was a shield. The same scale of penalization was for mistakes made while weighing gold, with a range of half zhu , i.e., 0.3 gram and above.

Capacity Apparatus. For tong (桶, same as hu 斛), the range for mistakes was set with 2 sheng (升) and above, the official would be penalized for the equivalent of a set of armor; between 1 and 2 sheng, the penalty was a shield. For dou (斗), the range for mistakes was set with 1/2 sheng and above, the official would be penalized for the equivalent of a shield, and so on.

The methods taken in that period had tremendous influences on the following feudal societies for more than two thousand years.

皮盾牌
Leather shields

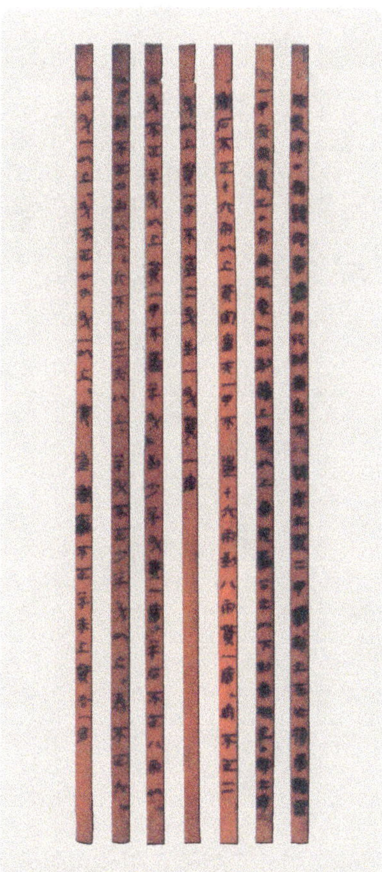

墨书度量衡检定、处罚条例的秦律竹简
Bamboo slips,written with ink on legal regulations set in the Qin Dynasty on verifications of measures and weights and on penalty

身着铠甲的武士
A warrior in armors

THE HISTORY  OF ANCIENT CHINES

EASURES AND WEIGHTS

6

THE HISTORY OF ANCIENT CHINESE MEASURES AND WEIGHTS

中国古代计量史

六、承上启下的汉代度量衡

Chapter Six

Measures and Weights in the Han Dynasty

汉代度量衡是在继承前代的基础上发展起来的。无论在标准的建立、单位制的制定，还是在器具的制造等方面都取得了很高的成就，为我国古代度量衡奠定了基础。

“汉承秦制”是指刘邦建立西汉王朝一直到汉宣帝，在一段相当长的时期继承和发展秦朝各项制度，度量衡也不例外。汉兴以后，刘邦即令张苍根据秦制“定度量衡程式”，使汉代度量衡制度很快地建立起来。从出土的秦汉度量衡器来分析研究，明显看到汉代度量衡基本上继承了秦制。由于秦立国时间甚短，虽然秦始皇在统一度量衡时已形成了一套上下关联的单位制，但却并没有系统地整理成文。

西汉末年，王莽篡权。为政治上的需要，王莽征集了当时学识渊博、通晓天文与乐律的学者百余人，考证前代制度，其中包括度量衡。经过总结归纳，整理成审度、嘉量、权衡各篇，后收入《汉书·律历志》。该书首次明确规定度量衡以黄钟为标准，以及度量衡单位名称、进位关系和与其相适应的标准器的制造等。该书成为我国古代度量衡史上最系统、最权威的著作。

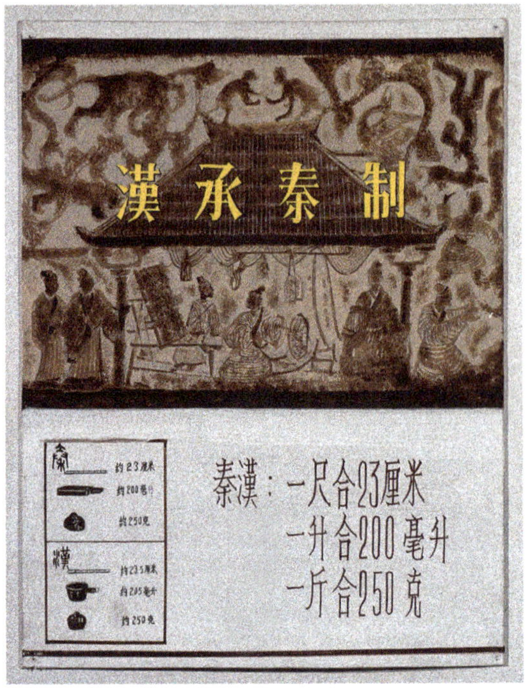

秦汉：一尺合 23 厘米，一升合 200 毫升，一斤合 250 克
The Qin and Han dynasty, 1 chi=23 cm, 1 sheng=200 ml, 1 jin=250 g

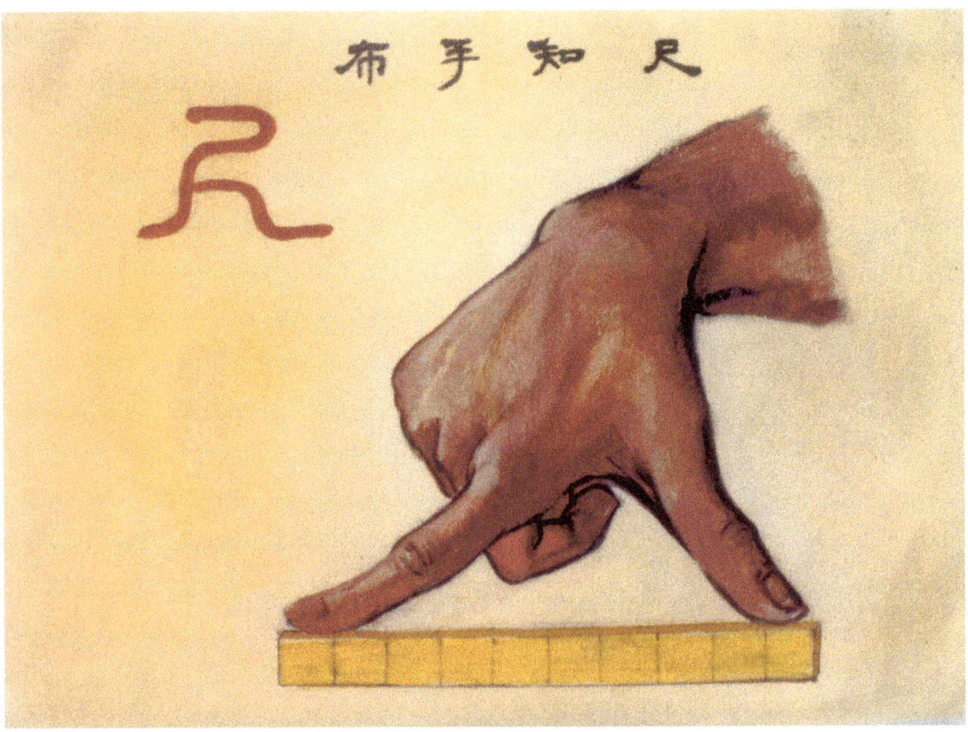

《孔子家语》记：“布指知寸，布手知尺”
In *Confucius Domestic Affairs*, a statement read “measuring cun with fingers,and measuring chi with the span”

On the bases of achievements made in the previous dynasties, the weighing and measuring system was progressing rapidly in the Han Dynasty (汉代). The facets of development ranged from the establishment of standards and units to the production of apparatus, which laid the foundation for a complete ancient Chinese weighing and measuring system.

Institutions of the Qin Dynasty were inherited in the Han Dynasty. During a long period of time from Liu Bang (刘邦) who established the Western Han (西汉) to the time of Xuan Di (宣帝), institutions of the Qin Dynasty had been inherited and carried forward, including the weighing and measuring system. In the Han Dynasty, shortly after Liu Bang mounting the throne, he ordered Zhang Cang (张苍) to "fix the system of measuring and weighing" referring to the system of the Qin Dynasty. Unearthed measuring and weighing apparatus made in the Qin and the Han dynasties reveal that the Han Dynasty inherited those of the Qin Dynasty. Although there existed a coherent system of units, the short duration of the Qin Dynasty saw it an impossible task for scholars of that time to write up the system and present the complete structures.

At the end of the Western Han, Wang Mang (王莽) seized power. In an effort to meet his political demands, he summoned hundreds of learned scholars who were knowledgeable about temperament and astronomy together to examine and correct measuring and weighing systems in the past dynasties. They classified those materials in the old times and put them in new categories such as Inspections of Measures (审度), Customization of Capacity (嘉量), and Regularity of Weights (权衡), which were complied in the Han Shu / Lu Li Zhi (《汉书·律历志》). These were the most authoritative works for the first time in the Chinese history to specify systematically the standards, units, production of apparatus, and so forth.

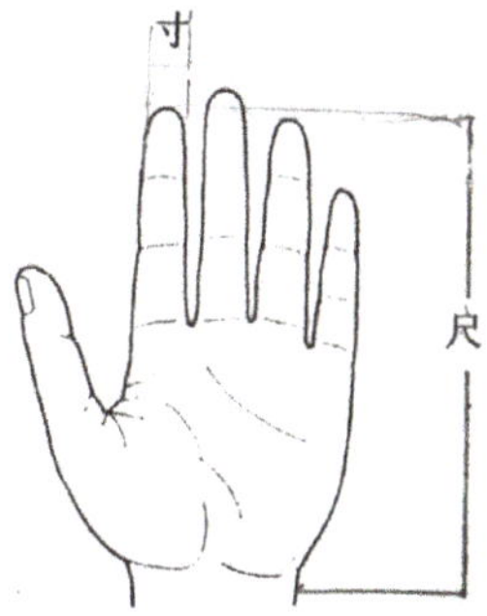

指宽与手之比约为 1:10
The ratio between the fingerbreadth and the hand is approximately 1:10

《小尔雅》记："跬，一举足也，倍跬为步。"240 方步为一亩
In *Xiao Er Ya*, it read, "Kui is the length of 1 step,and bu is the length of 2 steps." 1 mu is the area of 240 square bu

1. 标准的建立

（1）用人体定度量衡标准

人们早期的测量常常借用人体来测定一个量，如“布手知尺”“迈步定亩”“手捧成升”等。由于成年人的身体各部位不会有太大的区别，在建造原始半地穴式房屋或分配谷物时，用这些方法进行测量，其精度尽管粗疏，但只要在使用它的一定范围内具有通用性，也就达到目的了。随着人类生产技术的提高，如果没有一个统一的标准，测量时出现较大误差而反映出来的矛盾便日益突出，并且成为影响准确测量的障碍时，就需要建立相对准确并具有权威性的标准。从而出现了禹“身为度”，即将大禹身体某一部分的长度定为一个长度单位，并且制造出专用的测量器具，这一器具便成为治水工程中唯一的标准了。

图间插文 Descriptions matching the Pictures

原始时的度量衡往往以人体来定出一个量。《史记》有大禹治水“身为度，称以出”的记述。世界上许多国家都有类似的记载。

In the primitive times, units of measures and weights were set according to the lengths of human bodies. In *Shi Ji*, there was a record that Da Yu, while regulating rivers and floods, set measures and weights on the basis of his height and weight. There exist lots of similar records in other countries all over the world.

1. The Establishment of Standards

(1) Setting the Standards According to the Bodily Measures

In the primitive times, people often took bodily measures as standards, and such methods included application of hands and feet, as in “measuring chi (尺) with spans”, “setting mu (亩) with steps”, and “handful capacity for sheng (升)”, etc. As there were no noticeable differences in the physical sizes among adults, such rough measuring and weighing methods were acceptable for measuring the sizes of shelters and for weighing grain. The inaccuracy became obvious with the development of production and technology, thus calling for the standardization of measuring and weighing systems that bore authority. An instance was Yu (禹) who took his height as the measurement, namely, physically, his height or the length of a part of his body was taken as the unit of length. Apparatus were made accordingly after these lengths which were in use for the hydro projects under his leadership.

《小尔雅》记：“一手之盛谓之溢，两手谓之掬，掬，一升也。”两手之盛与古一升容量约略相当

In *Xiao Er Ya*, it read, “Yi is the amount of 1 handful; Ju is 2 times of yi, equal to 1 sheng.” Hence ju was roughly the capacity of 1 sheng in the ancient times

(2)用自然物定度量衡标准

我国古代为建立度量衡标准，始终做着不懈的努力，并且积累了丰富的经验。出土于殷墟的三支骨尺，长度皆在16厘米左右，正好是中等身高人拇指尖至食指尖间一庹的长度，与“布手知尺”相吻合。古人深知用人体作长度标准误差很大的道理，因此一直在寻找一种复现性较好的自然物来定义一尺的长度。《淮南子·天文训》中说：“十二粟而当一寸。”《说文》：“十发为程，十程为分。”《孙子算经》：“蚕吐丝为忽，十忽为秒，十秒为毫，十毫为厘，十厘为分。”说法各异，然而都难以与汉代尺度相符。《汉书·律历志》记：“以子谷秬黍中者，一黍之广度之……一为一分。”黍和粟都是我国古代主要的粮食作物。粟即小米，古代又称稷。黍子多为椭圆形，平滑而有光泽，有白色、黄色和褐色多种。《汉书·律历志》所说的“一黍之广度之”，即选用中等大小的黍横排，一粒黍的宽度相当于一分之长。为求证此说，我们曾做过多次试验：用山西、北京等地所产浅黄色和深褐色的黍子横排100粒，约合23厘米，与汉代一尺之长约略相合。在2000多年前的秦汉时期，应用数理统计法求得平均值来定一尺之长，是科学的。比起“布手知尺”的复现性有很大提高，准确度也基本上能满足当时日常使用的要求。

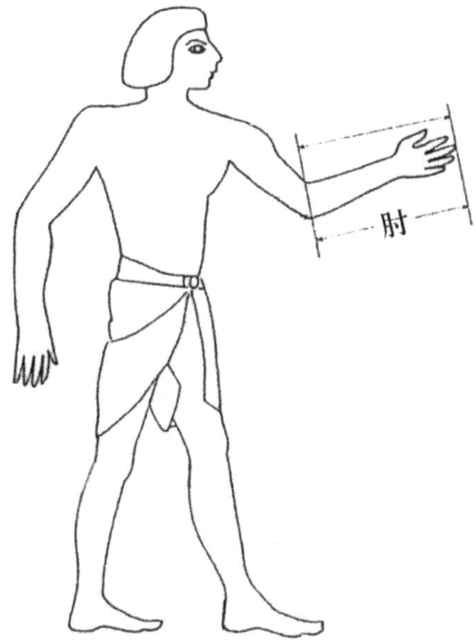

古埃及 以指尖至肘的间距定作一个长度单位，称肘尺或腕尺
In ancient Egypt, the length between the finger tips and the elbow was set as a unit which was called zhou chi (elbow ruler) or wan chi (wrist ruler)

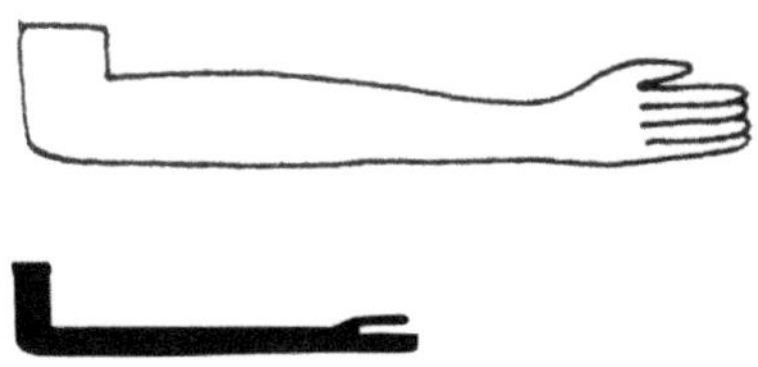

肘尺的象形字
Pictographs of an elbow ruler

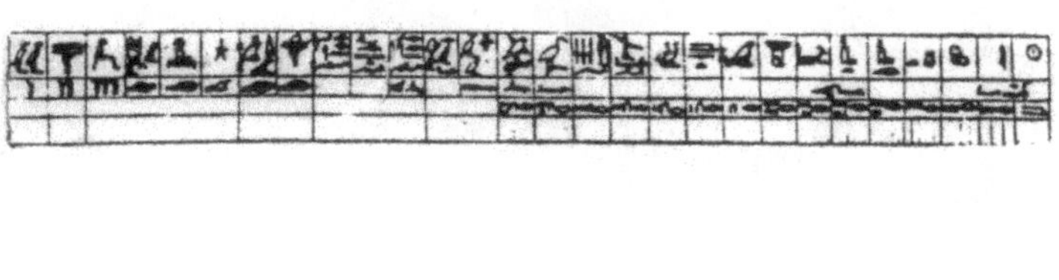

公元前约1500年埃及使用的两种肘尺，长约50厘米
The two types of elbow rulers in ancient Egypt, c. 1500 B.C. The length was about 50 cm each

(2) Using Natural Objects as Standards

Ancient Chinese people tried continuously to find applicable standards for the measurement, and accumulated many experiences in this process. It was said that three gu chi (骨尺 bone rulers) were unearthed from the Yin Ruins (殷墟), and each had a length of 16 cm, which was equal to 1 tuo (庹, span) of a man of medium height. This verified the old words saying "measuring chi with spans"(布手知尺). They were aware that discrepancies would occur when using human body as standards, so they never stopped trying to find natural objects with precision to define the Chinese mathematical foot, which serve as the basic unit of the linear measurement. In *Huai Nan Zi / Tian Wen Xun* (《淮南子・天文训》), it is said that 1 cun (寸) was equal to the total length of 12 millets(粟). In *Shuo Wen* (《说文》), "10 times of the diameter of a piece of hair on the ponytail is equal to 1 cheng (程), and 10 cheng was equal to 1 fen (分)." In *Sun Zi Suan Jing* (《孙子算经》), "the length of silk from a silk worm at one time represents 1 hu (忽). 10 hu was equal to 1 miao (秒), 10 miao was equal to 1 hao (毫), 10 hao was equal to 1 li (厘), 10 li was equal to 1 fen (分)." However, all these definitions varied from one another, and none of them coincided with the prevailing measures at that time. Relatively more authority was found in the definitions recorded in *Han Shu / Lv Li Zhi* (《汉书・律历志》), "the width of a corn millet (黍) represents 1 fen (分)".Both millet and corn millet were among the main crops in ancient China. Millet was also called ji (稷) in ancient China. Corn millet is a kind of crop shaped in ellipse and colored in yellow, white or brown. In order to prove the definition of the Chinese mathematical foot recorded in *Han Shu / Lv Li Zhi* , we made experimentations trying to line up 100 corn millets grown in Beijing (北京) and Shanxi (山西) areas side by side, then we found that the length was around 23cm, which was just the length of 1 chi (尺) in the Han Dynasty (汉代). In the Qin Dynasty (秦代) and Han Dynasty (汉代), to define the chi (尺) with the average diameter of corn improved the precision and accuracy of measurements. It was more reliable than the previous "measuring chi with spans". This chi (尺), as a basic unit of measurement, could meet requirements of daily life at that time.

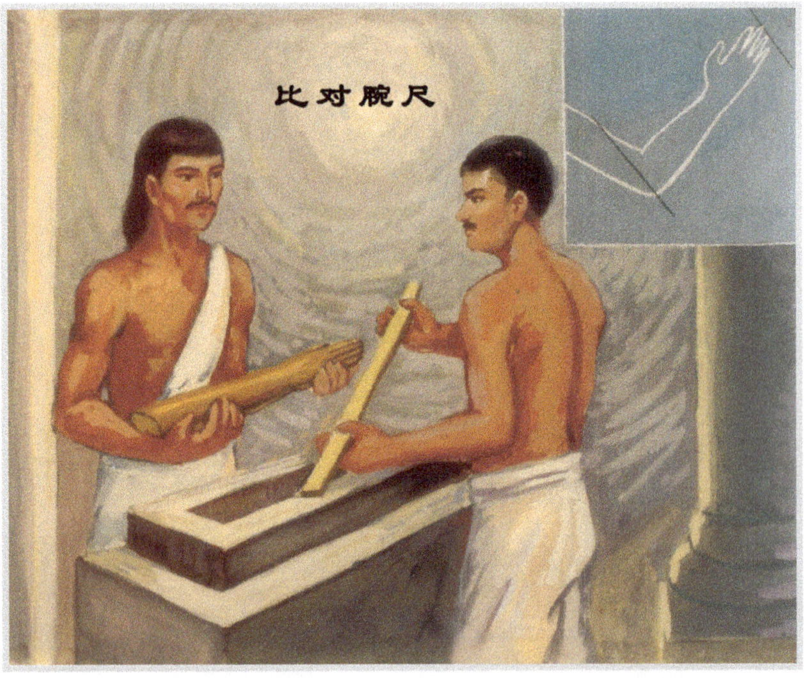

腕尺比对图
Elbow rulers in comparison

(3) 用有固定音高的黄钟律管定度量衡标准

用发出固定音高的黄钟律管定长度标准，是中国古代伟大的发明创造。这种方法与几千年后的现代，曾采用光波之长作为长度基准，其基本理论有惊人的相似之处。

中国古代未把度量衡列为专门学科，而多与音律、历算学并存，尤其是与音律学互为参证，成为古代度量衡史上的一大特点。由于在古代中国还没有频率的概念，乐律学家通常用尺来量度弦或律管的长短以确定音高，这就是古代度量衡与乐律学之间有着密切关系的缘由。

律管所发出的声音是由律管本身的固有频率所决定的，如果管径不变的话，频率和管长就有确定的关系：管子长，声音就低；频率增加一倍，音调也就提高一个八度。因此，把管子的口径和长度确定下来，那么这支律管所发出的绝对音高也就被确定下来。故要做一支具有一定频率的律管，它的长度和管径也应是确定的。在古代，如果要确定黄钟律管的音高，只能用耳朵去分辨。一位好的乐工能准确地判断出各种音调。历史上关于用耳来分辨乐律的记载屡见，如西晋时乐律学家荀勖，奉命以古律重整当时的宫廷乐律。荀勖凭着他敏锐的听力指出，与古乐相比，当朝的黄钟调声低了，于是重新制作了一支黄钟律管。随即用这支律管去校尺，果然发现比古尺长了四分有余(0.92 厘米)。后来经过考证，荀勖所定的尺确实与新莽时期的尺度基本相符。证明古代乐律学家能把握住黄钟律，就可以依据律管的长度而复制出标准的尺度了。然而，"量"可以载于实体，"声"却看不见、抓不住，只能凭耳朵

1531 年德国《几何学》中的木刻图，绘出 16 人挤在一起，后一人脚尖碰前一人脚跟，定出"一杆"的长度

A woodcut in German *Geometry* (1531), showing that the length unit of 1 gan(杆) was set by measuring the length between the tiptoes of the last person and the heels of the first person of a crowd of 16 standing in a line

英国曾将亨利一世（1100 ～ 1135 在位）鼻至指之间的距离定为一个长度单位

In Great Britain, the length between his nose and finger tips of Henry I (reigned 1100-1135) was set as a unit of length

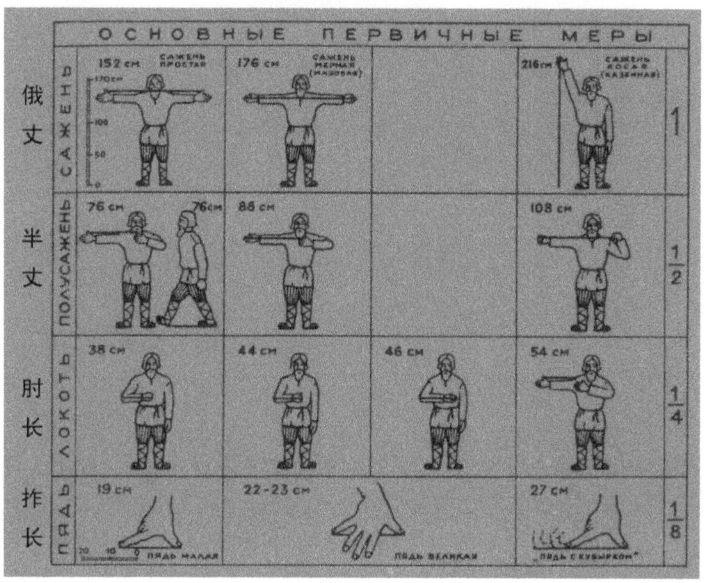

古俄罗斯国用人体各部位的比例定出长度单位，分"肘长"和"拃长"等

In ancient Russia, the units of lengths were set on the basis of scales of various parts of the human body, such as elbow lengths and hand spans

听。所以古人说："音失之甚易，求之甚难。"要想把度量衡三个量的实体，寓意于无形的声和数之中，毕竟还是难以复现的，于是便有了累黍定律之法。《汉书·律历志》是最早把乐律与度量衡的关系用累黍的方法具体化的古代文献。《汉书·律历志》记载："度者……本起黄钟之长，以子谷秬黍中者，一黍之广度之，九十分黄钟之长，一为一分……""量者……本起于黄钟之龠，用度数审其容，以子谷秬黍中者，千有二百实其龠……""权者……本起于黄钟之重，一龠容千二百黍，重十二铢……"上述三段话的主要内容是，用累黍的方法，加上与度量衡三个量相互参照，从而把律管长度、管径确定下来。这种把无形的乐律与具象的实体相结合的做法，是古人经过反复试验的总结，也是汉代在度量衡史上一项重大创造性的发明。经过我们反复验证，用累黍的方法来校量度量衡，是能够达到相对准确度的。我们的做法是：选用1200粒中等大小的黄褐色秬黍，第一步是横排100粒，得到长度约23厘米，与新莽时期的尺基本吻合；第二步用1200粒黍测其容，约11.5毫升，略大于新莽时期一龠之量（莽时一龠当为10毫升）；第三步再用这1200粒黍测其重，得7.4克，与新莽时期12铢之重相近。

尽管无论是用累黍定尺，还是用黄钟律管定尺，都还存在一系列难于确定的技术规范。但是如果我们考虑到当时的具体条件，而以累黍定尺与布手知尺来比较，前者的精确度毕竟高了很多。在汉代，利用金属和纯水相对密度确定度量衡单位量值标准的方法，也已得到应用。《汉书·食货志》："黄金方寸，而重一斤。"《后汉书·礼仪志》："水一升，冬重十三两。"

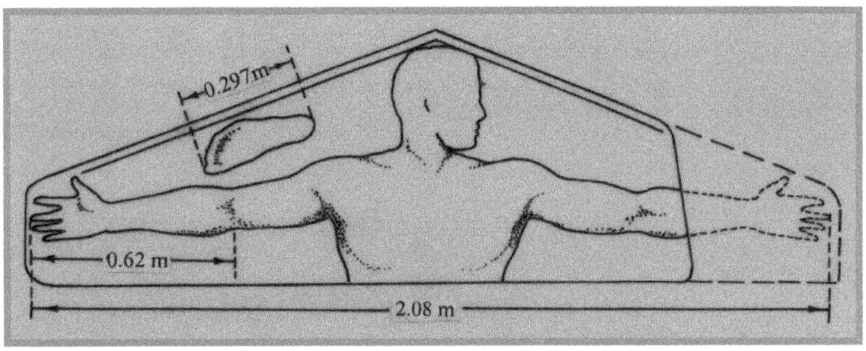

公元450年英国将人体长度的比例刻在浮雕上
Rilievo showing the scales of the parts of the human body, in Great Britain in 450

德国将最先走出教堂的16名男子脚长的1/16，定为一个长度单位
In Germany, 1/16 of the foot of the 16 men who were among the first stepping out of the church was set as a unit of length

(3) Using Bamboo Pitch-Pipes to Customize the Standards of Measures and Weights

To fix the primary standards with the Bamboo Pitch-Pipes (黄钟律管) was a great invention of the ancient Chinese people. This method was similar with fixing the primary standard with the value of light-wave in modern times. The tone of a pitch pipe was determined by the frequency the pipe itself produced. When the diameter of the pipe was fixed, the correlation between the length of the pipe and the frequency was thus fixed. The longer the pipe, the lower the pitch would be. And when the frequency was increased, the pitch was raised accordingly. In the ancient times, the pitch of a pitch pipe could only be distinguished by human ears. Ancient China was a nation of music and rites. Most of scholars at that time were good at music and highly sensitive to the variations of pitch. A talented professional musician was able to tell different pitches precisely. All these can be seen in ancient books and records. For example, in the Western Jin (西晋) times, a scholar on temperament named Xun Xu (荀勖) was ordered to standardize royal music of his time according to ancient one. He found with his ears that the pitch of music of his time was generally lower than that of the old times. So he made a standard pitch pipe coinciding with old ones. Afterwards, he used the newly made pitch pipe to check against the former standard rods, finding that they were 4 fen (分 0.92cm) longer than the old ones. It was then proved that the foot reset by Xun Xu was just the same as that fixed in the Xin Mang (新莽) Period. This example tells us that so long as some scholars on temperament were able to distinguish the pitch precisely, then standard rods would be made in accordance with the length of pitch pipes. Nevertheless, "quantity" is an entity while "tone" is invisible that can only be caught by ear. Hence ancient people commented that "it is easy to lose sounds and difficult to catch". In order to embody the entities of measures and weights in the invisible sounds and numbers, *the Law of Lei Shu* (累黍) came into being, which was a method of determination by way of accumulating and counting the number of pieces of corn millet. *Han Shu / Lv Li Zhi* (《汉书・律历志》) was the earliest historic document recording the correlations between musical temperaments and measuring and weighing by using *the Law of Lei Shu*

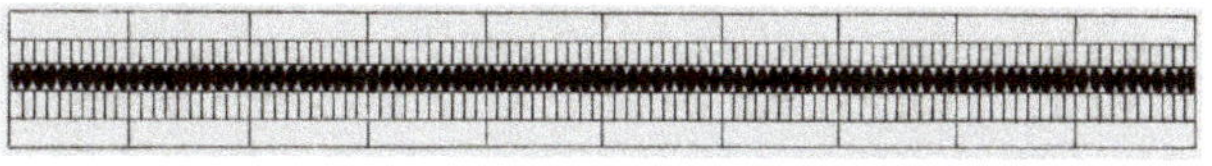

《汉书・律历志》记："以一黍之广度之……一为一分。"横排 100 粒黍，正合汉代一尺之长，约 23 厘米

In *Han Shu / Lv Li Zhi*, it was recorded that "the width of a corn millet represents 1 fen". 100 pieces of corn millet aligned side by side represented 1 chi in the Han Dynasty, about 23 cm

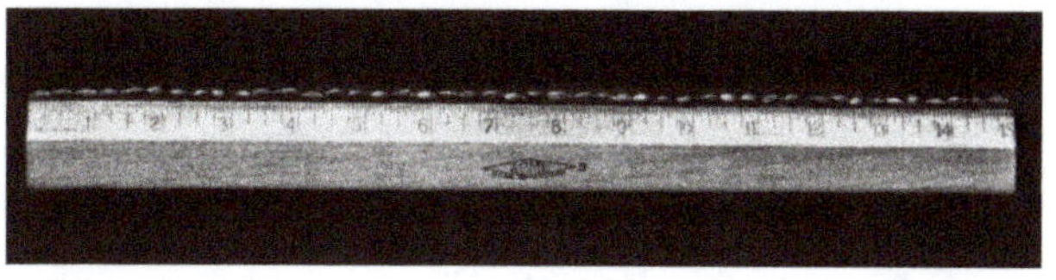

《英国度量衡史》记：以 45 粒大麦首尾相连排列，定为 15 英寸

In *The History of the British Measuring and Weighing Systems*, it is recorded that the length of 45 pieces of aligned barley was set as 15 inches

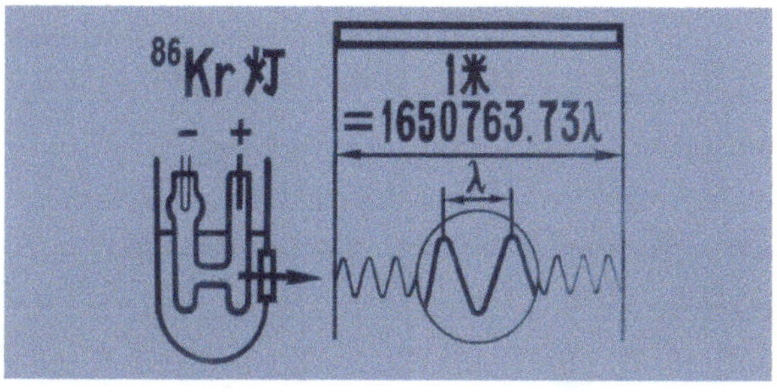

1960 年第 11 届国际计量大会决定，改用氪-86 在特定条件下发出的橙黄色光波波长的 1650763.73 倍的长度为 1 米

In 1960, at the Eleventh International Plenary Session of Metrology, it was decided to adopt 1,650,763.73 times of the length of orange-colored light wave of 86kr, under a specially designated circumstance, as 1 meter

《汉书·律历志》记度量衡以"黄钟"为本，即以发出固定音高，长度合 9 寸的黄钟律管定长度标准。此中蕴含着用声波定长度并以黍为介质，导出容量和重量的单位量

In *Han Shu / Lv Li Zhi*, it was recorded that measures and weights were set in accordance with "the Bamboo Pitch-Pipes (Huang Zhong)", about 9 cun in length with a fixed pitch. The implication was that lengths were set in accordance with sound waves; and corn millet was applied as a medium, thus reaching units of capacity and weight

(累黍). *Han Shu / Lv Li Zhi* (《汉书·律历志》) started with "coinciding music temperaments with measuring and weighing, for the goal of unity across the land and credibility among the subjects", and "length is in accordance with the measure of huangzhong (黄钟), in reference to the length of a piece of ju shu (秬黍,corn millet)". "Capacity... was originally stemming from the unit yue (龠) of huangzhong in reference to the measurement of a piece of corn millet". "Weighing was also stemming from the weight of huangzhong in reference of yue that contained 1,200 pieces of corn millet, and weighed 12 zhu (铢)". The contents of the above quotations indicate that the length and diameter of a pitch pipe was determined with the help of the method of accumulating and counting the number of pieces of corn millet,with the values of measuring and weighing taken into consideration as reference. Such an invention of combining the invisible music with substantial entities was of great significance in the history of measuring and weighing in the Han Dynasty (汉代). It was the result of numerous experiments by ancient people. Based on our own experiments by adopting the same method, using *the Law of Lei Shu* , we also found that accuracy was reached. The following demonstrates the way we did out experiments. Selecting and gathering 1,200 pieces of medium sized brown corn millet.

Step 1. Arranging 100 pieces side by side, and the length read 23 cm, which was the about the length of chi (尺) in the Xin Mang (新莽) Period.

Step 2. Checking the capacity of the 1,200 pieces, that read 11.5ml, which was a little bit greater than that of yue (龠) in the Xin Mang Period (1 yue was about 10 ml at that time).

Step 3. Weighing the 1,200 pieces, that read

公元 1791 年法国规定了以通过巴黎的子午线，从赤道到北极的一千万分之一为 1 米

In 1791, it was regulated in France that one ten millionth of the meridian across Paris from the equator to the North Pole was set as 1 meter

出土的汉代 12 支律管(明器)

12 unearthed pitch-pipes in the Han Dynasty (for the nether world)

7.4g, which was approximately the same as 12 zhu (铢) in the Xin Mang (新莽) Period. It has to be pointed out that owing to the lack of technical regularities at that time, either *the Law of Lei Shu* , or the pitch pipe as the standard, was not a perfect method to determine the length of chi (尺). However, compared with the previous method of "measuring chi with spans", greater accuracy and precision of measurement was reached by adopting *the Law of Lei Shu* and the standard of pitch pipe. In the Han Dynasty (汉代), Chinese people started to apply the mass of metal and water to setting the standards for the unit value in the measuring and weighing system. In *Han Shu / Shi Huo Zhi* (《汉书・食货志》),it was recorded that "a gold bullion with the volume of 1 cubic cun (寸) weights 1 jin (斤)". In *Hou Han Shu / Li Yi Zhi* (《后汉书・礼仪志》), it was recorded that "1 liter of water weights 13 liang (两) in winter".

2. 单位制的制定和标准器的制造

《汉书·律历志》中对度量衡三个量与黄钟、秬黍的关系，各个量的单位名称、进位制以及标准器的制作都有详细的规定。近年来，新莽时期制造的度量衡标准器多有发现，可与《汉书·律历志》中审度、嘉量、权衡各篇相互印证。为了便于对汉代度量衡有一个比较全面的了解，我们将这时期单位制和标准器按审度、嘉量、权衡三部分来叙述。

(1)审度

《汉书·律历志》云："度者，分、寸、尺、丈、引也，所以度长短也。本起黄钟之长。以子谷秬黍中者，一黍之广度之，九十分黄钟之长。一为一分，十分为寸，十寸为尺，十尺为丈，十丈为引，而五度审矣。其法用铜，高一寸，广二寸，长一丈，而分寸尺丈存焉。用竹为引，高一分，广六分，长十丈。……职在内官，廷尉掌之。"

1926年在甘肃省定西县出土了一批新莽时期的度量衡器，其中有一支铜丈，上刻王莽统一度量衡81字诏文："黄帝初祖，德匝于虞，虞帝始祖，德匝于新，岁在大梁，龙集戊辰，戊辰直定，天命有民，据土德受，正号即真，改正建丑，长寿隆崇。同律度量衡，稽当前人，龙在已巳，岁次实沉，初班天下，万国永遵，子子孙孙，亨传亿年。"大意是：初祖黄帝，始祖虞帝，皆为先祖。斗转星移，周而复始至初始元年(9)。诏告万民：王莽于摄政三年后登天子之位。将初始元年十二月改为始建国元年正月。以黄钟乐律考正度量衡，合于虞帝之始。始建国元年颁度量衡于全国。这一制度将由子孙后代传至万世，永远遵从。铜丈未刻分寸线纹，只有长、宽、高端面。实测铜丈长229.3厘米，宽4.7厘米，厚2.4厘米，与《汉书·律历志》所述铜丈高一寸，广二寸，长一丈的比例基本相

合。铜丈只有端面而无分、寸、尺的线纹，故并非实用器，而是作为校准用的标准器。

目前所见保存在全国各地的两汉尺近 100 支，半数以上为铜质。除铜尺以外，还有铁尺、骨尺、牙尺、竹尺和木尺等。尺上或以各种纹饰分成 10 个寸格，或以线纹分割成 10 寸，每寸刻 10 分。两汉尺度基本沿用秦制，每尺在 23 厘米左右，东汉后期尺度略有增长，每尺约长 23.5 厘米。

除一般的直尺外，汉代已有了专用的测长工具——卡尺。 尺正面刻有铭文 "始建国元年正月癸酉朔日制"，铭文表明此尺制作于新莽始建国元年。卡尺由固定尺和滑动尺两部分组成，两端均有成矩形的量爪。固定尺正面刻 40 分格(即四寸)，上部有一鱼鳞形柄，中间开一导槽，滑动尺正面刻有 5 个寸格（未刻分)，量爪与尺身相联处有环状拉手，引环可使滑动尺移动，当两尺的量爪靠拢时，固定尺与滑动尺等长，两尺刻线大体相对。

卡尺的发明解决了测量工件外圆直径尺寸的问题。而在此之前，凡涉及外圆直径，常常用围长代替。卡尺除了便于用量爪来测量圆径、板厚的尺寸外，还可以用固定尺的右端面为基面，引其环移动活动尺，使活动尺右端面与槽底面接触，便可测得槽深尺寸了。

新莽卡尺的发现，说明距今 2000 多年前的西汉末年，我国测量长度的技术已从一支直尺发展到能制造出既可以测量直径，又便于测量深度的多用途的工具。卡尺的发明是长度测量技术上的一个突破。

所见汉代尺多达 90 余支，材质、纹饰不尽相同，长短略有参差，但都在 23 ～ 23.5 厘米

90 rulers made in the Han Dynasty, with various types of materials and grains, and different lengths between 23 cm and 23.5 cm

2. The Determination of Units and the Production of Standard Devices

In *Han Shu / Lv Li Zhi* (《汉书·律历志》) the relationships between the value of units in the weighing and measuring system with the pitch-pipe and corn millet were described. The regularities were given to the names of those values, the calculation of measuring or weighing units and their fractions, and to the technologies of producing standard measuring and weighing devices. Recently many measuring and weighing devices made in the Xin Mang (新莽) Period were unearthed. Those devices can be reciprocally verified with what were described in *Han Shu / Lv Li Zhi* as Inspections of Measures (审度), Customization of Capacity (嘉量), and Regularity of Weights (权衡).In order to obtain a comprehensive understanding about the measuring and weighing system in the Han Dynasty (汉代), here we introduce the system of units and standard devices in the following three sections of Inspections of Measures, Customization of Capacity, and Regularity of Weights.

(1) Inspections of Measures (审度)

In *Han Shu / Lv Li Zhi* it was recorded that du (度), literally meaning "measuring length", comprised fen (分), cun (寸), chi (尺), zhang (丈) and yin (引). Hence du was used to measure lengths of objects. The du was in reference of the linear measurement of huangzhong (黄钟, the Pitch Pipe) and corn millet. The array of 5 units in such linear measurement using lei shu (累黍) was as follows—10 fen (分) = 1 cun (寸), 10 cun = 1 chi (尺), 10 chi = 1 zhang (丈), 10 zhang = 1 yin (引).The devices used for inspecting the measures were bronze yardstick and bamboo yardstick. The dimensions of the bronze yardstick—Height: 1 cun. Width: 2 cun. Length: 1 zhang. The dimensions of the bamboo yardstick—Height: 1 fen. Width: 6 fen. Length: 10 zhang. The governmental official assigned to be in charge of the inspection of measures was entitled Ting Wei (廷尉).

In 1926, some measuring and weighing devices made in the Xin Mang (新莽) Period were unearthed in Dingxi (定西) County, Gansu (甘肃) Province. Among those devices there was a bronze yardstick (铜丈) on which

新莽时期 始建国元年(9)制造的专用测长工具——卡尺
A caliper made in the Xin Mang Period, used especially for measuring lengths

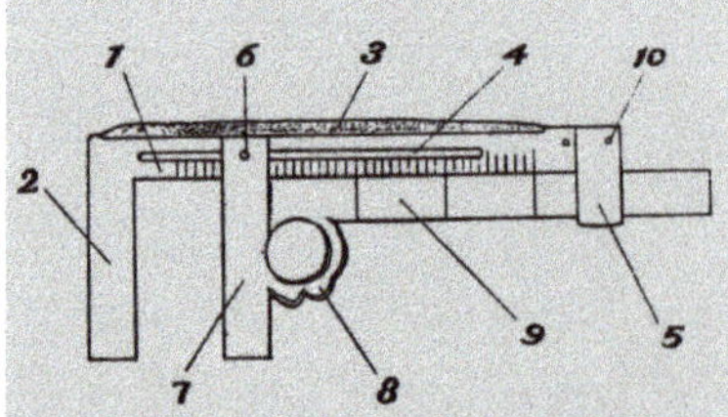

1. 固定式
2. 固定卡爪
3. 鱼形柄
4. 导槽
5. 组合套
6. 导销
7. 活动卡爪
8. 拉手
9. 活动尺
10. 铆钉

铜卡尺结构图
A diagram of the structure of a bronze caliper

1992 年东汉墓出土的铜卡尺
A bronze caliper made in the Eastern Han Dynasty, unearthed in 1992

engraved the order by the Emperor Wang Mang (王莽) to standardize the system of weights and measures. The engraved order by the Emperor also indicated the significance of implementing such standardization: It was from the providence inherited by the former emperors, starting from Huang Di (黄帝), for the benefits of all subjects a generation after another forever. Theoretically, we can calculate precisely the actual value of 1 chi (尺) in the Han Dynasty (汉代) by comparing the recorded length, width and height of the yardstick with the real measuring results. Since the results of the actual dimensions were 229.3 cm in length, 4.7 cm in width and 2.4 cm in height, which coincided with what was recorded in *Han Shu / Lv Li Zhi* (《汉书・律历志》) about the measures of the bronze yardstick : 1 cun (寸) in height, 2 cun in width, and 1 zhang (丈) in length. Since there is no graduation on this yardstick, we can conclude that it is not a yardstick for practical use but for calibration. That is to say, it is a device meant for inspection of measures.

The total of unearthed yardsticks/rulers made in the Han Dynasty is around 100. Half of them are made of bronze, and the rest of them are made of iron, bone, animal tooth, bamboo or wood. One chi in length, there were measures carved on each of these rulers with a scale of 10 cun marked by decorative or geometrical patterns, and each cun was further divided by patterns into 10 fen (分). The linear measurement in the Han Dynasty inherited that in the Qin Dynasty (秦代). As the fundamental unit, chi in the Han Dynasty and in the Qin Dynasty was almost identical, measured around 23cm. In the late Eastern Han Dynasty, the chi was set a little longer, measured around 23.5 cm.

古代工匠使用卡尺图
A sketch of craftsmen using a calipers in the ancient times

Calipers (卡尺) were also invented in the Han Dynasty. On the surface of these two calipers there engraved characters indicating the date and year when it was made— the first year of the Xin Mang (新莽) Period. Each of these two calipers was composed of a fixed ruler and a movable one. The fixed ruler carried a scale of 40 fen equal to 4 cun , while the movable ruler carved with 5 cun scales. This invention made it convenient to measure the diameters of round objects, which were previously calculated by measuring girths. Besides, calipers were also used to measure the depth of troughs. Such invention over 2,000 years ago was a great breakthrough in the history of measuring technology.

斛

斗

升

(2) 嘉量

《汉书 · 律历志》说："量者，龠、合、升、斗、斛也，所以量多少也。本起于黄钟之龠，用度数审其容。以子谷秬黍中者，千有二百实其龠，以井水准其概。十龠为合，十合为升，十升为斗，十斗为斛，而五量嘉矣。其法用铜，方尺而圜其外，旁有庣焉。其上为斛，其下为斗，左耳为升，右耳为合、龠……职在太仓，大司农掌之。"

"新莽嘉量"是一件五量合一的铜质标准量器。量器的主体部分是一个大圆柱体，近下端有挡隔，挡隔上方是斛量，下方是斗量；左侧是一个小圆柱体，为升量；右侧也是一个小圆柱体，挡隔在中端，上为合量，下为龠量。故斛、升、合三量口朝上，斗、龠二量口朝下。器壁正面有 81 字总铭，与新莽铜丈铭文相同。单件量器上还各有分铭，记录了各量的尺寸和容积。嘉量设计巧妙，制作精湛，刻铭详尽，又与文献互为佐证。历史上许多律、算学家如刘徽、祖冲之、李淳风等都对它做过考校。

近人刘复将新嘉量的尺寸、容量、重量做了精密的测量，著有《新嘉量之校量及推算》。根据实测，推算出新莽时期一尺长 23.1 厘米，一升容 200 毫升，一斤重 226.7 克。"新莽嘉量"辗转流传至今，是研究两汉度量衡的珍贵文物。

除"新莽嘉量"外，所见两汉量器数十件，包括龠、合、升、斗、斛等各种量值。质地大多数为铜，其中许多镌刻铭文来说明它的用途、量值和使用方法。1953 年甘肃省古浪县出土了一件"大司农平斛"。腹壁刻铭文一行："大司农平斛，建武十一年(35)正月造。"大司农是汉代官职，掌管国家赋税钱谷盐铁等主要财政收入，也主管量器的制造、检定与发放等工作。"平斛"是指公平、均等的意思，有如今天的标准量器。"大司农平斛"是由大司农监制、校量的"斛"，即国家级的标准量器。经实测，铜斛容 19600 毫升，折合 1 升为 196 毫升。今搜集到两汉时的量器皆经实测容量，折算 1 升均在 200 毫升左右。证明了汉代沿用秦之遗制。

汉代 小铜量容 6 毫升
A small bronze instrument in the Han Dynasty with a capacity of 6 ml

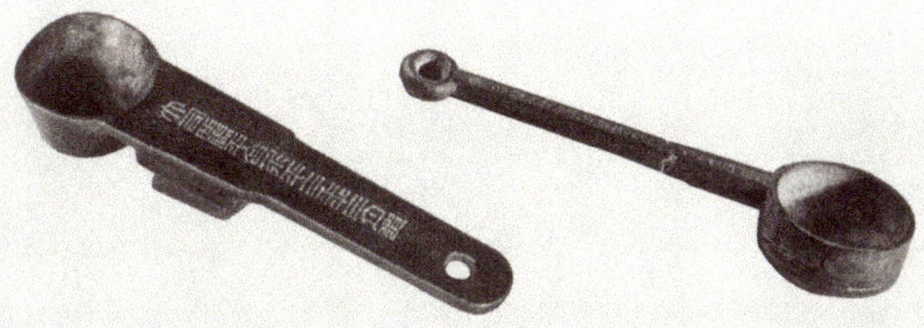

合　　　龠

汉代的容量单位制：2 龠 1 合、10 合 1 升、10 升 1 斗、10 斗 1 斛。1 升约容 200 毫升，是秦制的沿袭。近代出土汉代量器多件，量值基本准确

Units of capacity in Han Dynasty: 2 yue = 1 ge, 10 ge = 1 sheng, 10 sheng = 1 dou, and 10 dou = 1 hu. 1 sheng was about 200 ml, which was inherited from the systems prevailing in the Qin times. The values of capacities of numerous weighing instruments in the Han Dynasty unearthed recently were accurate

新莽时期　铜嘉量及铭文拓本　嘉量是公元 9 年制造的标准量器。全器包括了龠、合、升、斗、斛五个量，与《汉书·律历志》记载相对照，不仅可以详尽地了解汉代的容量制度，而且通过对器物铭文的研究和对量器的测量，还可以得出度量衡三者的单位量值。1 尺合 23.1 厘米，1 升容 200 毫升，1 斤重 226.7 克

Rubbings of the inscriptions on the Bronze (Tong) Jia Liang used as a set of standard weighing instruments, made in the Xin Mang Period, A.D. 9, which consisted of 5 capacity units of yue, ge, sheng, dou and hu. This set is not only in reference with what was recorded in *Han Shu / Lv Li Zhi*, but also serves for research to find out the values of the units of length, capacity and weight. 1 chi was about 23.1 cm, 1 sheng was about 200 ml, and 1 jin was about 226.7 g

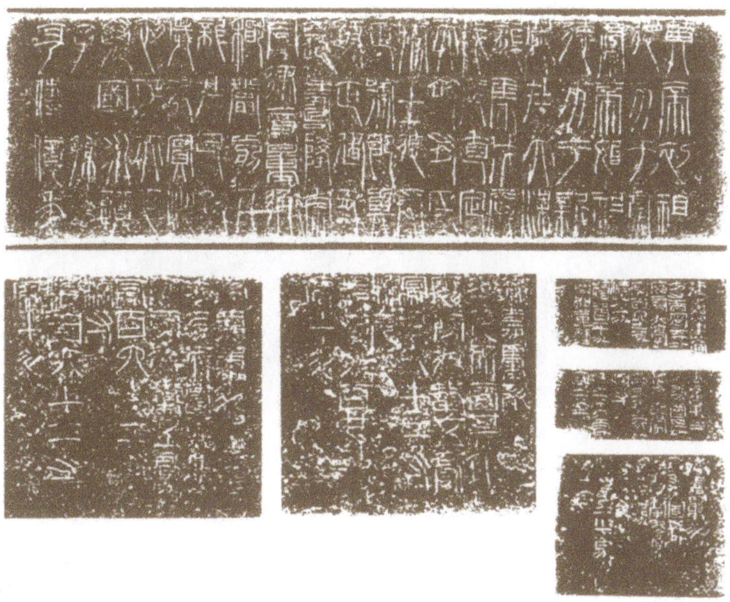

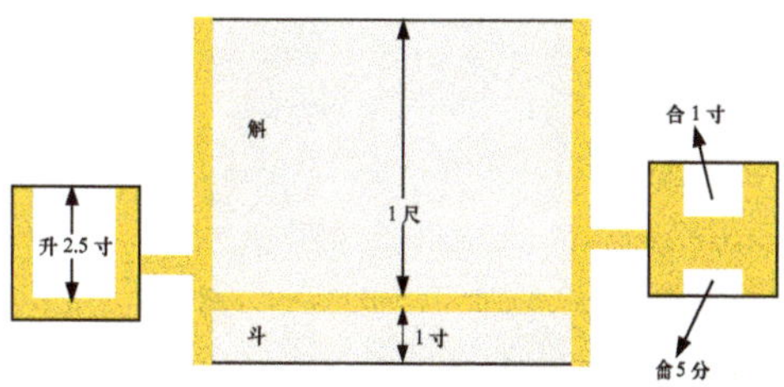

新莽时期 铜方斗 方斗口沿有铭文:"律量斗,方六寸,深四寸五分,积百六十二寸,容十升。始建国元年(9)制"

A bronze fang dou of the Xin Mang Period, made in A.D. 9, with inscriptions on the rim that read, "Standard instrument for capacity, 6 cun by side, 4 cun and 5 fen in depth, 162 cubic cun, and 10 sheng in capacity

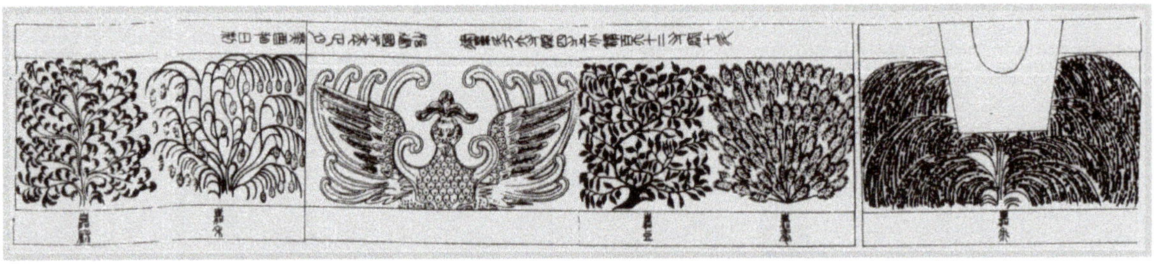

(2) Customization of Capacity (嘉量)

In *Han Shu / Lv Li Zhi* (《汉书·律历志》) it was recorded that capacity, liang (量) comprised yue (龠), ge (合), sheng (升), dou (斗), and hu (斛). Hence liang was meant as du (度) for measuring capacity, namely, the amount of objects. This measurement was stemming from the yue of huangzhong (黄钟), in reference with lei shu (累黍) wherein the capacity of 1,200 pieces of corn millet was measured as 1 yue . There were five scaled units (五量) in this array— 10 yue = 1 ge , 10 ge = 1 sheng , 10 sheng = 1 dou , 10 dou = 1 hu. The measuring device was made of bronze with carved lines indicating on the top as hu , the bottom as dou , left side as sheng, right side as ge and yue. The governmental official in charge of this customization of capacity was entitled Tai Cang (太仓) and Da Si Nong (大司农).

Xin Mang Jia Liang (新莽嘉量) is a bronze standard device of capacity. It contains all the five scaled units. On the main body of the device hu was carved at the bottom, dou at the lower part, sheng on a smaller column-shaped device annexed to the left side of the main body, and on another smaller column-shaped device annexed to the right side, ge was carved on the upper part, and yue on the lower part. On the front side the name and time of the device were carved, which were the same as bronze yardstick (铜丈) made in the Xin Mang (新莽) Period to record the units of length and capacity. The device of customization of capacity demonstrates a delicate design and artifacts with detailed carved characters. As it was also verified by historic records, there had been many scientists and mathematicians like Liu Hui (刘徽), Zu Chongzhi (祖冲之) and Li Chunfeng (李淳风) who conducted researches and examinations on the device. Liu Fu (刘复) of the modern times had a

新莽时期 统一度量衡铜诏版及拓本
Rubbings of an imperial edict stating the unified measuring and weighing systems in the Xin Mang Period

precise measurement of the Cross-Checking and Calculation on the Xin Jia Liang (新嘉量), from its size and capacity to weight, and described the results in his book entitled *Cross-Checking and Calculation of Xin Jia Liang* (《新嘉量之校量及推算》). Liu's findings of the measurements in the Xin Mang (新莽) Period included the following: 1 chi (尺) = 23.1 cm, 1 sheng (升) = 200 ml, and 1 jin (斤) = 226.7 g. The availability of Xin Mang Jia Liang (新莽嘉量) at the present times serves as a valuable object from the ancient times for researches on the measuring and weighing systems in the Western and Eastern Han dynasties.

Apart from the Xin Mang Jia Liang , there are also several score of pieces of devices for capacity whose measuring units include yue (龠), ge (合), sheng (升), dou (斗), and hu (斛). They were mostly made of bronze, with descriptions of their purposes, capacities, and instructions on how to use them.In 1953, Da Si Nong Ping Hu (大司农平斛) was unearthed in Gulang (古浪) County, Gansu (甘肃) Province. Judging by the characters Da Si Nong (大司农) engraved on the surface, one could tell that it was produced in the Han Dynasty (汉代). Da Si Nong was the title of an official in the Han Dynasty, in charge of the production, inspection and distribution of measuring devices, as well as tax levying and other financial matters (equivalent to the minister of finance in the modern times). The characters Ping Hu (平斛) meant "fair-ness" and "equality", indicating that the main function of this object was serving as the

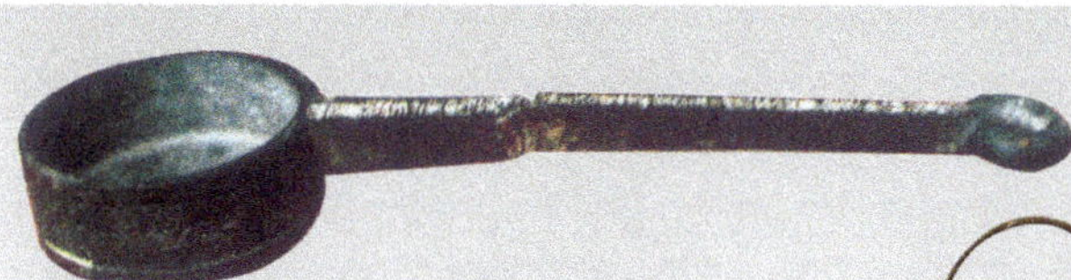

新莽时期 铜龠 柄上刻铭文:"始建国元年正月癸酉朔日制""律量龠,方寸而圜其外,庣旁九豪,冥百六十二分,深五分,积八百一十分,容如黄钟"。容 10 毫升,计算铜龠尺长 23.03 厘米
A bronze yue with inscriptions on the handle that read, "Produced in the first year of the Xin Mang Period", "Standard capacity for yue, square shape inside and round shape outside; mouth area of 162 square fen, 5 fen in depth, 810 fen in volume, like that of huangzhong". The capacity was 10 ml, and the length was 23.03 cm

standard device of capacity. Because Da Si Nong Ping Hu (大司农平斛) was produced under the supervision and inspection of Da Si Nong (大司农), a ministerial official in the central government, it was the standard device of capacity at the national level. After measuring, it was found that the actual capacity of this bronze device could hold 19,600 ml, with 196 ml equal to 1 sheng (升). Based on the results of examining this type of devices unearthed and collected at present, and of cross-checking the actual capacities of them, it has been proved that the devices for measuring capacities made in the Western and Eastern Han dynasties were following the standards stipulated in the previous Qin Dynasty (秦代) when 1 sheng (升) was roughly equal to 200 ml.

新莽时期 铜斛 刻新莽统一度量衡诏书

An imperial edict carved on a bronze hu stating the unified measuring and weighing systems in the Xin Mang Period

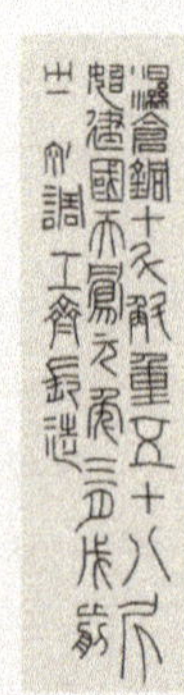

新莽时期 "漯仓"铜斛 此为地方仓廪使用的铜斛，容 19100 毫升

Luo Cang Bronze Hu in the New Mang Period used in local granaries with a capacity of 19,100 ml

西汉时期“常(尚)方”铜量 自铭:“容半(斗),重五斤。”容1020毫升,重1240克

Shang Fang Bronze Instrument (Tong Liang) made in the Western Han Dynasty with inscriptions as "capacity of half dou, weighing 5 jin". The capacity is 1,020 ml, and weight 1,240 g

新莽时期“夷道官”铜斛 由地方根据国家标准复制的铜斛。容20300毫升

"Yi Dao Guan" Bronze Hu in the Xin Mang Period with a capacity of 20,300 ml, an instrument duplicated by local authorities according to the state standard

东汉时期“都市平”铜量 地方“市”上交易用的铜升

Du Shi Ping Bronze Instrument made in the Eastern Han Dynasty, an instrument used on local markets

东汉时期“市平斗”铜量

Shi Ping Dou Bronze Instrument made in the Eastern Han Dynasty

东汉建武十一年(35)“大司农”铜斛 容19600毫升
Da Si Nong Bronze Hu with a capacity of 19,600 ml, made in the 11th year of Jian Wu (A.D. 35) in the Eastern Han Dynasty

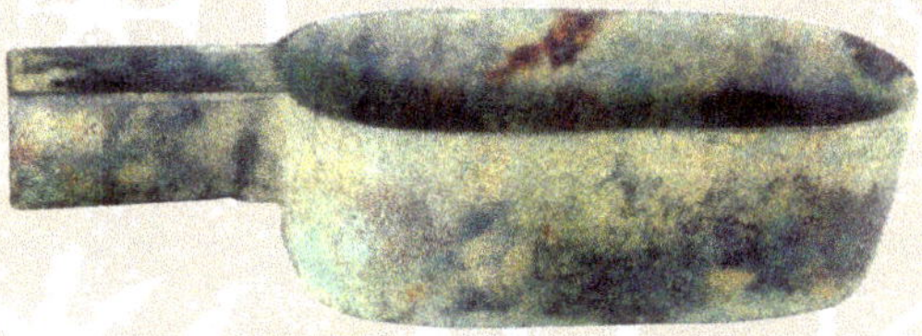

西汉时期“杨氏”铜量 自铭重“八斤十二两”。容1960毫升，正合一斗量
Yang Shi Bronze Instrument made in the Western Han Dynasty, with inscriptions as “weight 8 jin and 12 liang”. The capacity was 1,960 ml, which was just 1 dou

新莽时期“大半龠”铜量 容6.7毫升
Da Ban Yue (greater than 1/2 yue) Bronze Instrument of the Xin Mang Period with a capacity of 6.7 ml

西汉时期“平都”铜量 自铭“容三升少半升，重二斤十五两”。容645毫升
Ping Du Bronze Instrument made in the Western Han Dynasty, with inscriptions as “capacity of 3 sheng and less than 1/2 sheng, weight 2 jin and 15 liang”. The capacity was 645 ml

东汉光和二年(179)“大司农”铜斛 口、底沿皆刻相同的 98 字铭文，大意是：大司农依照《黄钟律历》《九章算术》制造了度量衡器发至各州，并规定每年秋分之日进行检定，以保证全国统一。容 20400 毫升

Da Si Nong Bronze Hu made in the 2nd year of Guang He (A.D. 179) in the Eastern Han Dynasty, with a capacity of 20,400 ml. There are inscriptions both at the rim and the bottom, 98 characters each, with the main points as Da Si Nong inspected the production of this measuring and weighing instrument in accordance with *Huang Zhong Lv Li* and *Jiu Zhang Suan Shu*. Such instruments were distributed to all the local zhou (an ancient administrative division). It was stipulated that the instruments must be examined on the day of autumnal equinox, to ensure that the use of such instrument was unified nationally

一组小铜量
A series of bronze instrtruments for capacities

东汉时期 一组石质量器 容量 200 ~ 2500 毫升
A series of stone instruments made in the Eastern Han Dynasty, with capacities varying from 200 ml to 2,500 ml

(3) 权衡

《汉书 · 律历志》云："权者，铢、两、斤、钧、石也,所以称物平施，知轻重也。本起于黄钟之重。一龠容千二百黍，重十二铢，两之为两。二十四铢为两，十六两为斤，三十斤为钧，四钧为石。圆而环之，令之肉倍好者……职在大行，鸿胪掌之。"

1926 年在甘肃省定西县秤钩驿与新莽铜丈同时出土了新莽时期制造的铜衡杆、铜钩和多件铜环权。衡杆中间为提纽，两端各有扁平长方体的悬纽，衡杆中部刻新莽统一度量衡诏文 20 行共 81 字，与新莽铜丈、新莽嘉量总铭相同。衡杆重量合新莽衡制 10 斤。同时出土的还有石铜权、二钧铜权、九斤铜权、六斤铜权、三斤铜权各一件。铜权为扁平环状，外径约为孔径的 3 倍。按各件权的自铭折算，1 斤在 238 克左右。

此外，实测汉权共 45 件(套)，器形分鼻纽形和圆环形两种。权衡是古代称重器的统称，直至近现代才分天平、台秤、木杆秤等多种称谓。经过对各方面材料的综合分析，我们认为秦和西汉时的鼻纽权，尽管形制类似今天杆秤上用的秤砣，但使用方法上则是提纽在中间而权和被称物置于衡杆的两端，称重时以权的已知重量来求被称物的未知量，相当于今天的等臂式天平及砝码。西汉之后，权的形制虽然没有大的变化，但使用上有的已转变为不等臂的杆秤秤砣。杆秤的量程广，使用和携带方便，制作也简单，受到广大群众的欢迎，汉代以后广泛使用并流传至今。

纵观我国度量衡发展史，秦汉时期无论是在标准的确立、器物的制作，还是在单位制的完备、制度的建立等方面，都占有重要的地位。此后虽因王朝兴衰、改朝换代，常常要重新颁定度量衡制度，但是每每以追求秦汉古制为准则。2000 多年来，度量衡无论是量值的增大或减小、器具的兴废或更换、单位制的改进或复古，其基本体制始终是秦汉时期所建立的。鸦片战争以后，外来制度不断输入，直到 1984 年法定计量单位公布和实施，才彻底地摒弃了旧制，推行米制。因此，可以说秦汉时期是我国度量衡发展史上一个重要的时代，对后世影响也极为深远。

新莽时期 铜衡杆、铜钩、铜环权 衡杆上刻新莽统一度量衡诏书，重 2442 克。铜环权 5 件：石铜权(重 29950 克)、2 钧铜权(重 14775 克)、9 斤铜权(重 2222.8 克)、6 斤铜权(重 1446.2 克)、3 斤铜权(重 730 克)

Bronze Balance Beam, Hook and Hoop. Made in the Xin Mang Period. An imperial edict was carved on the beam indicating the unified measuring and weighing system at that time. The beam was 2,442 g in weight. There were 5 pieces of weights in hoop shape: 1 stone weight (29,950 g in weight), 2-jin weight (14,775 g), 9-jin weight (2,222.8 g), 6-jin weight (1,446.2 g) and 3-jin weight (730 g)

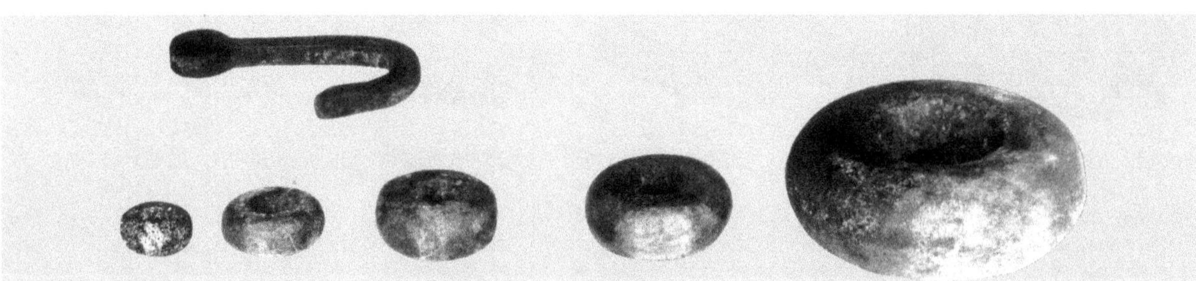

西汉时期 铁权 铸铭文"十五斤"。重 3575 克
An iron weight made in the Western Han Dynasty cast with inscriptions of "15 jin". It is 3,575 g in weight

新莽时期 铜环权 权身刻自重铭文"律权石，重四钧"。重 29950 克
A Bronze Hoop Weight in the Xin Mang times, with inscriptions of its weight as "Standard Weight for Dan, 4 Jun in Weight". It is 29950 g in weight

(3) Regularity of Weights (权衡)

In *Han Shu / Lv Li Zhi* (《汉书 · 律历志》) it was recorded that weight(quan , 权) included zhu (铢), liang (两), jin (斤), jun (钧), and dan (石). Such array of 5 quan (权) as a set of scales was stemmed from the weight determined in huangzhong (黄钟) where—

1 yue (龠) = 1,200 corn millet, weighing 12 zhu .24 zhu = 1 liang ,16 liang = 1 jin , 30 jin = 1 jun ,4 jun = 1 dan .The governmental official in charge of this kind of affairs was entitled Da Xing (大行 upright and influential conduct), Hong Lu (鸿胪).

In 1926, a good number of weighing devices were unearthed in Dingxi (定西) County, Gansu (甘肃) Province. These devices were divided into two types: bronze balancing rod (铜衡杆) and bronze circle-shaped quan (铜环权), all were made in the Xin Mang (新莽) Period with descriptive characters similar with those on the bronze yardstick (铜丈) and Xin Mang Jia Liang (新莽嘉量). Of the bronze balancing rod, the heaviest weighing device, a balance rod, weighed 10 jin itself. There were other weighing devices of the same type, weighing 9 jin, 6 jin , and 3 jin respectively. Of the bronze circle-shaped quan, they were flat with the diameter of the hole 1/3 of that of the external diameter. Carved on each of the objects was the respective weight, and it was calculated that 1 jin (斤) was equal to 238 g.

Quan Heng (权衡) was a general term in ancient times, referring to weighing devices for capacities. There appeared divisions with new terms in the contemporary times, such as "balance", "scale", and wooden cheng (木杆秤). Based on synthesized analyses, it has been found out that the principles of using quan in weighing objects in the Qin and Western Han dynasties were similar to that of using balances in modern times.After the Western Han, wooden cheng became more and more popular because it was easier to produce and more convenient to use. It is still

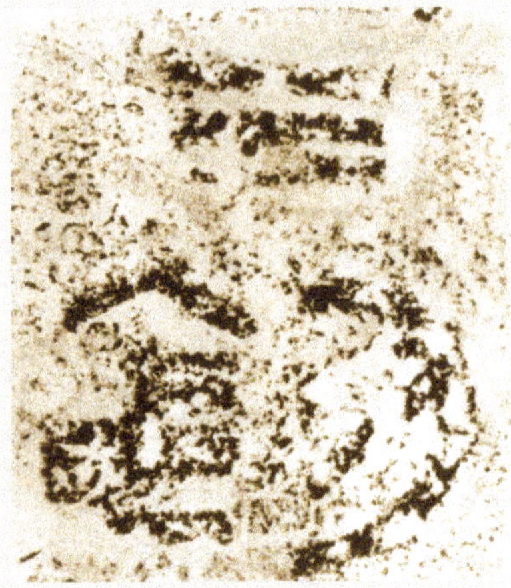

西汉时期 铁权 铸"三钧"。重 22490 克
An iron weight made in the Western Han Dynasty, 22, 490 g in weight, with inscriptions of "3 jun"

being used at present.

A longitudinal revisit shows that what had been established in the Qin and Han times, from the determination of standards, production of devices to the setting of units and stipulation of the system, played an important role throughout the history of development of the measuring and weighing system in China. Even with frequent changes of dynasties throughout the history, the systems set in the Qin and Han times were referred to whenever a new system was to set in a new dynasty. For over 2,000 years, the basic measuring and weighing system was originally from the Qin and Han times, even there were changes in capacities, in application or abolishing the use of some devices, or in unit setting. It was the Opium War with influx of Western systems influencing the long-lasting measuring and weighing system, and the Law of Units on Measuring and Weighing Units stipulated in 1984 that put the old system to an end and started the implementation of the new metric system. Therefore, it can be claimed that the Qin and Han times were a period in the history of the development of the measuring and weighing system in China and that there have been farreaching influences throughout the history.

西汉时期 铜权 刻"武库一斤"。重 252 克
A Tong Quan (Bronze Weight), 252 g in weight, made in the Western Han Dynasty with inscriptions of "Wu Ku, 1 jin in weight"

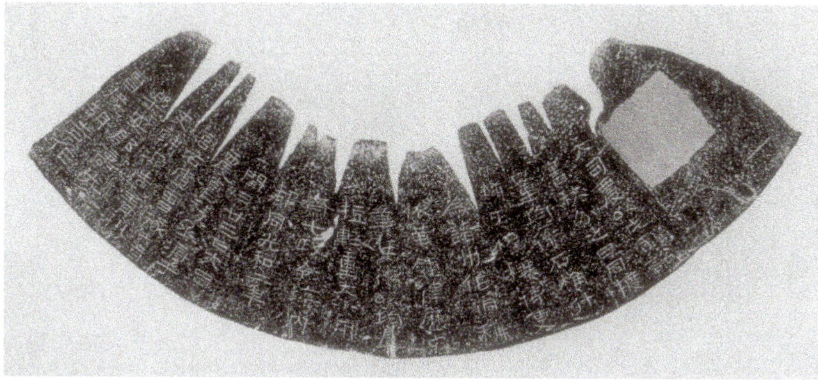

东汉时期 大司农铜权 刻有“大司农以戊寅诏书:秋分之日,同度量,均衡石,桶斗桶,特为诸州作铜秤……令海内都同”,诏文内容与光和二年(179)大司农铜斛基本相同。重 2996 克

Da Si Nong Bronze Weight, 2,996 g in weight, made in the Eastern Han Dynasty, with inscriptions of “Da Si Nong implementing the Wu Yin Imperial Edict on the day of autumnal equinox, for the purpose of ensuring that the units of measuring, weighing, and capacity were unified across the land”. The contents of the edict were about the same on the Da Si Nong Bronze Hu in the 2nd year of Guang He

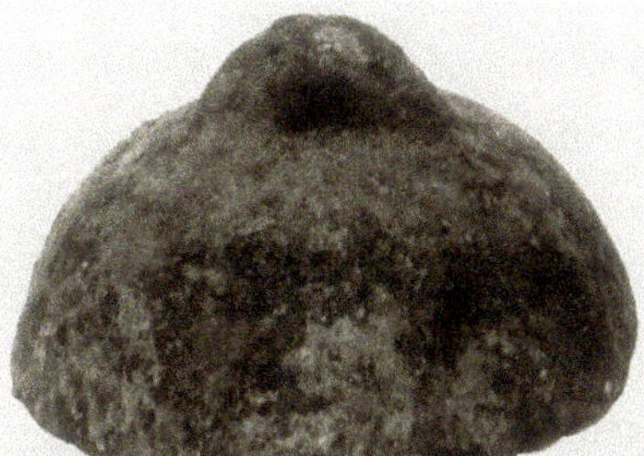

西汉时期 铁权 重 737 克,约合 3 斤

An iron weight made in the Western Han Dynasty, 737 g in weight, about 3 jin

西汉时期 官累铜权 刻“官累重斤十两”。“累”与权同义。重 400 克

A Guan Lei (Standard Weight) Tong Quan (Bronze Weight), 400 g in weight, made in the Western Han Dynasty with inscriptions of “Guan Lei, 1 Jin and 10 Liang in Weight”. “Lei” was the synonym of Quan(weight)

西汉时期 官累铜权 刻“官累重斤十两”。重 397 克

A Guan Lei (Standard Weight) Tong Quan (Bronze Weight), 397 g in weight, made in the Western Han Dynasty with inscriptions as “Guan Lei, 1 Jin and 10 Liang in Weight”

东汉时期 这几件各地出土的东汉铜、铁质权,单位量值都不呈倍比关系

Several unearthed bronze and iron weights made in the Eastern Han Dynasty whose values of weights were not in the relationship of times

东汉时期 铁权(8件):分别重736克、695克、229克、224克、216克、206克、134克、111克

8 iron weighs made in the Eastern Han Dynasty with weights of 736 g, 695 g, 229 g, 224 g, 216 g, 206 g, 134 g and 111 g respectively

图间插文 Descriptions matching the Pictures

"权衡"是中国古代称重器的统称。杆秤出现在什么时代尚无确证,但最晚在东汉时期已普遍使用了。

"Quan Heng" (Scale) was a general designation used for weighing objects in ancient China. Although there was no definite evidence as to when the gan cheng (steelyard) was first used, it became a common instrument for weighing as early as in the Eastern Han Dynasty.

秦汉的权衡器,如果皆用250克作为一斤的量值,那么东汉以前的权,虽然形制和质地不尽相同,但基本上皆可按整数倍来折算出每一件属于哪一级量值。因此它们多是在等臂衡杆(天平)上作砝码用。而东汉及其后的时代,情况则发生变化。无论是多件同时出土,也无论是鼻纽权或环形权、铜质权还是铁质权,往往很少能与某一数值形成倍比关系。因此判断它们已多数是在不等臂杆秤上作秤砣使用了。

With regard to the weighing instruments popular in the Qin and Han times, the weights (no matter what they were made of, or what shapes they were in) prior to the Eastern Han Dynasty, with 250 g as the value of 1 jin, could be used for weighing and calculating objects in round figures and times, hence they were used on the beams of scales/steelyards. Circumstances changed since the Eastern Han Dynasty. It was noticed that the weights unearthed, regardless of their shapes, and the materials they were made of, i.e., copper or iron, were unable to have a relationship of times with certain numbers. Hence it is judged that they were used as sliding weights on different types of steelyards.

秦代 丽山园铜缶 铭刻:"容十二斗三升,重二钧三斤八两"

A Li Shan Yuan Bronze Jar made in the State of Qin with inscriptions of "12 Dou and 3 Sheng in Capacity, and 2 Jun, 3 Jin and 8 Liang in Weight"

西汉时期 铜灯 铭刻："重三斤八两，高八寸"
A Bronze Lamp (Tong Deng) made in the Western Han Dynasty with inscriptions of "3 Jin and 8 Liang in Weight and 8 Cun in Height"

西汉时期 四连铜鼎 铭刻："容三斗，并重九十三斤"
A Si Lian Bronze Tripod (Tong Ding) made in the Western Han Dynasty with inscriptions of "3 Dou in Capacity and 93 Jin in Weight"

西汉时期 太后府铜钟 铭刻："容五斗，重十七斤"
A Tai Hou Fu Bronze Cup (Tong Zhong) made in the Western Han Dynasty with inscriptions of "5 Dou in Capacity and 17 Jin in Weight"

西汉时期 魏公铜壶 铭刻："容三斗二升"
A Wei Gong Bronze Kettle (Tong Hu) made in the Western Han Dynasty with inscriptions of "3 Dou and 2 Sheng in Capacity"

东汉时期 鎏金铜斛 铭刻："盘径二尺二寸"
A Gold-gilded Bronze Hu made in the Eastern Han Dynasty with inscriptions of "3 Chi and 2 Cun in Diameter"

图间插文 Descriptions matching the Pictures

历代青铜器上常见记自身尺寸、容量、重量的铭刻，汉代尤盛。
Inscriptions were often seen on bronze objects made in all the dynasties, about their sizes, capacities and weights. It was more prevailing in the Han Dynasty.

THE HISTORY  OF ANCIENT CHINESE

7

THE HISTORY OF ANCIENT CHINESE MEASURES AND WEIGHTS

中国古代计量史

EASURES AND WEIGHTS

七、从魏晋南北朝时期的混乱到隋代的统一

Chapter Seven

Transition From the Confusions in the Wei, Jin, Northern and Southern Dynasties to the Re-unification in the Sui Dynasty

3世纪，中国进入了三国时期(220～280)。这时期的特点是，统治阶级内部矛盾复杂、斗争激烈、战争频繁，社会长期处于分裂状态，维持了数十年的三国（魏、蜀、吴）鼎立局面，但度量衡制度却仍为汉制。

265年，司马炎改魏为晋，建立西晋(265～317)。280年，司马氏灭吴，结束了三国割据局面，统一中国。317年，西晋灭亡，司马睿在南方建立东晋(317～420)。东晋持续了100多年，而度量衡却未见有变制。

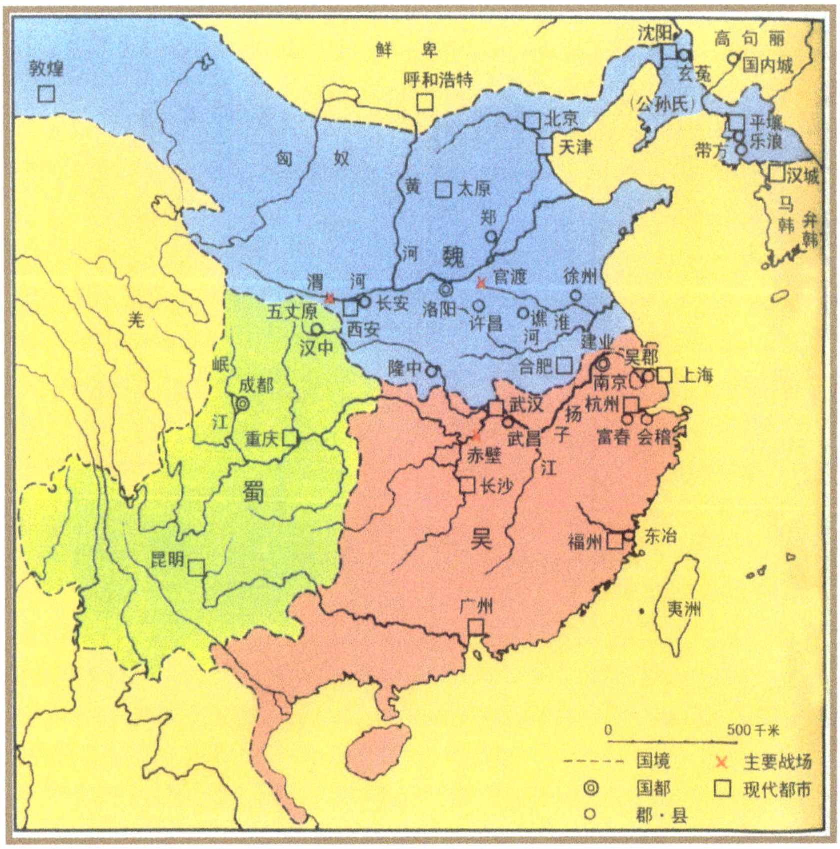

魏、蜀、吴三国鼎立形势图
A map of the Three Kingdoms of Wei, Shu and Wu in tripartite confrontation

During the Three Kingdoms (三国时期, A.D. 220-280) period, the Chinese society was characteristic of incessant warfare. Local war lords and powerful landlords in the three kingdoms were in control. In order to expand territories at each other's expense, they fought against each other or among themselves. But throughout the period of the Three Kingdoms (Wei,Shu and Wu)the measuring and weighing system carried in from the previous Han Dynasty remained in implementation.

In the year 265, Sima Yan replaced the Wei regime with his own, and named it as Jin (晋). Historians refer to this new regime as the Western Jin (西晋,A.D. 265-317). In the year 280, the Sima (司马) clan conquered Wu, ending the times of the Three Kingdoms and briefly unifying China. In the year 317, the year after the Western Jin was terminated, Sima Rui established a new regime in the south which historians refer to as the Eastern Jin (东晋,A.D. 317-420). The Eastern Jin had lasted for more than 100 years. Time changed, but no changes on the measuring and weighing system.

三国时期·魏 数学家刘徽考校秦汉度量衡制与魏制之比

Liu Hui, a mathematician in Wei of the Three Kingdoms, examining and comparing the measuring and weighing systems between Qin Han Dynasties and Wei

刘徽在《九章算术注》中说：新莽时期一斛当魏斛的九斗七升四合。折算得魏一斛容 20400 毫升

Liu Hui stated in *Jiu Zhang Suan Shu Zhu* that 1 hu in the Xin Mang Period was about 9 dou, 7 sheng and 4 ge in terms of hu in the Wei times. It is now computed and converted that the capacity of 1 hu in the Wei times was 20,400 ml

刘徽在《九章算术注》中说：魏尺比汉尺增长了四分七厘。折算得魏尺长 24.19 厘米

Liu Hui stated in *Jiu Zhang Suan Shu Zhu* that the length of 1 chi in the Wei times increased 4 fen and 7 li in comparison with that of chi in the Han Dynasty. It is calculated now that the length of 1 chi in the Wei times was 24.19 cm

1. 魏晋时期的度量衡

三国时期至晋代(西晋、东晋)度量衡,史书少有系统记述。留传下来的器物也比秦汉时大为减少。三国(魏、蜀、吴)政权是在东汉末年混乱中建立起来的,各种制度基本上沿袭汉制,度量衡也不例外。景元四年(263),数学家刘徽考证了新莽时的尺度,得新莽时一尺相当于魏尺的九寸五分五厘。这样便得知魏尺长度是24.19厘米。又据《晋书》记载,西晋时中书监荀勖受命定宫廷乐律。经多方考校,他认为三国时的尺度比新莽尺增长了四分七厘。他的考证与刘徽是一致的。经过我们的搜集测量,得到三国至晋尺共8支,长度在24.2～24.5厘米。文献记载与考古发掘所得实物互为佐证,也基本相同。故两晋一尺之长亦可厘定为24.2厘米。

除尺度之外,魏、晋时量器、权衡器均未见留存至今者。仅《晋书》记载了一件魏时的标准量器——大司农斛。经刘徽测算并与新莽嘉量作比较,可得知新莽时期一斛是魏斛的九斗七升四合。按前述,莽斛约合今20000毫升,那么便可以计算出魏时一斛约相当于今20400毫升。这是对两件标准量器作精确测量、计算和比较后的结果。这一微弱的差别,对古代一般的测量容积的器具来说,是可以忽略不计的。因此可以认为三国时的容量也是汉代制度的延续。由于迄今未见可供考校的晋代文献和实物,而从继其后的南朝(宋、齐、梁、陈)又多延续汉制来看,可以认为三国至晋代,量、衡也都是继承汉代制度的。

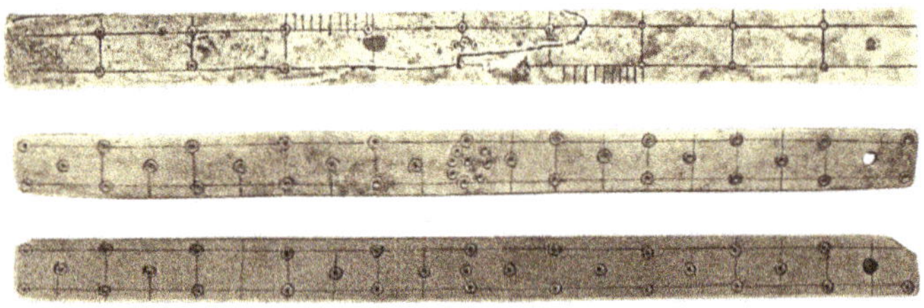

三国时期·魏 三支尺皆长23.8厘米
3 rulers of the Wei times in the Three Kingdoms, 23.8 cm in length each

西晋时期 骨尺 长24.5厘米
A bone ruler in the Western Jin times, 24.5 cm in length

1. The Measuring and Weighing System in the Wei and Jin Times

Systematical summarizations of the measuring and weighing systems of the Three Kingdoms and of the Western Jin and the Eastern Jin (西晋, 东晋) were seldom found in historical works. Unearthed measuring and weighing apparatus made in this period were much less compared with that of the Qin Dynasty. The Three Kingdoms, namely, Wei, Shu and Wu (魏, 蜀, 吴) rose from the turmoil of the end of the Eastern Han. Institutions of these three kingdoms were basically inherited from the Eastern Han, so was the measuring and weighing system. In the 4th year of Jingyuan (景元, 263), a mathematician of Wei whose name was Liu Hui (刘徽) made a textual research on the linear measurement of the Xin Mang (新莽) Period, and reached a conclusion that 1 chi of that time was equal to 9 cun (寸) 5 fen (分) and 5 li (厘). From this historical record we notice that 1 chi (尺) of Wei (魏) was 24.19 cm . In *Jin Shu* (《晋书》 *the History of Jin*) there are some materials related to the measurement of that time. In the Western Jin a man named Xun Xu (荀勖) who was Zhong Shu Jian (中书监), an official position mainly in charge of classified documents and sometimes with involvement in the government and political affairs, received instructions to determine the royal temperament (宫廷乐律). He conducted a research on the temperament of previous dynasties and drew a conclusion that the length of chi (尺) in the times of the Three Kingdoms was 4 fen (分) and 7 li (厘) longer than that in the Xin Mang Period. This conclusion coincided with what Liu Hui found out.It was recorded in many historical works that the length of chi (尺) in the Western and Eastern Jin (西晋, 东晋) times was the same as that in the Three Kingdoms. These records were proved by some results of archaeological studies. We collected and measured eight rulers made with different materials in the periods from the Three Kingdoms to Jin, and found that the lengths were between 24.2 cm and 24.5 cm, hence we concluded that the length of 1 chi (尺) in the Jin times was 24.2 cm.

Beside those unearthed rulers, the Wei and Jin times left no measuring apparatus for us to study. In *Jin Shu* (《晋书》) there were some materials about a capacity device called Da Si Nong Hu (大司农斛). It was indicated in this historical record that Liu Hui (刘徽) was measuring the capacity of this object and comparing the capacity with that of the Xin Mang Jia Liang (新莽嘉量). Liu Hui found that 1 hu (斛) in the Xin Mang (新莽) Period was 9 dou (斗), 7 sheng (升) and 4 ge (合) in Wei. Since the Mang Hu (莽斛) was around 20,000 ml, the Wei Hu (魏斛) should be 20,400 ml. Therefore we can roughly conclude that the capacity measuring system was inherited from the Han Dynasty. Because there is no materials of the Jin times related to the measuring and weighing system left for our study, and because the system in the Southern Dynasty (南朝, Song, Qi, Liang and Chen) was inherited from the Han Dynasty, we believe that from the period of the Three Kingdoms to Jin the measuring and weighting system basically kept the same as that in the Han Dynasty.

北朝时期·魏 铜砣 重 248 克

A bronze sliding weight in Wei of the Northern Dynasty, 248 g in weight

2. 南北朝时期的度量衡

南北朝(420～589)包括南朝的宋、齐、梁、陈和北朝的北魏、东魏、西魏、北齐和北周。南北朝时期度量衡有两大特点:①其间单位量值增长速度为历朝之冠。②制度混乱,各个时期、各个地区量值相差悬殊的现象十分突出。以尺度而论,唐代天文、数学家李淳风为求古律历,曾将汉至隋各种尺度以长短分别列为十五等,史家称之为“十五等尺”并多有考证。其中南朝宋时(420～479)尺长 24.5 厘米,比东晋约略增长 2%,仍是秦汉尺度的延续,并被齐、梁、陈所沿用。而北朝尺度则迅速增长,并且十分混乱。如北魏(386～534)前期一尺长 25.5 厘米,中期长 27.9 厘米,后期长 29.5 厘米。至东魏(534～550),一尺已长达 30 厘米。北齐(550～577)取代东魏,承袭其各种制度。纵观北朝 100 年中,尺度增长了约 6 厘米。

容量和权衡制度,南朝和北朝增长率也不同。南朝梁、陈两朝,升斗仍沿袭新莽旧制,每升约合 200 毫升。齐朝量制大于古升,约 300 毫升。北朝量制增长速度很快,北魏时每升“于古二而为一”,约 400 毫升,其后又有增长。至隋文帝统一全国,建立隋朝时,容量 1 升已增至约 600 毫升。

权衡制,南朝和北朝的单位量值相距也很大。南朝的梁、陈,1 斤约合 220 克,齐 1 斤约合 330 克。而北朝的增长速度却远远超过了南朝。据文献记载:“魏、齐斗秤于古二而为一”,推算 1 斤当合今制

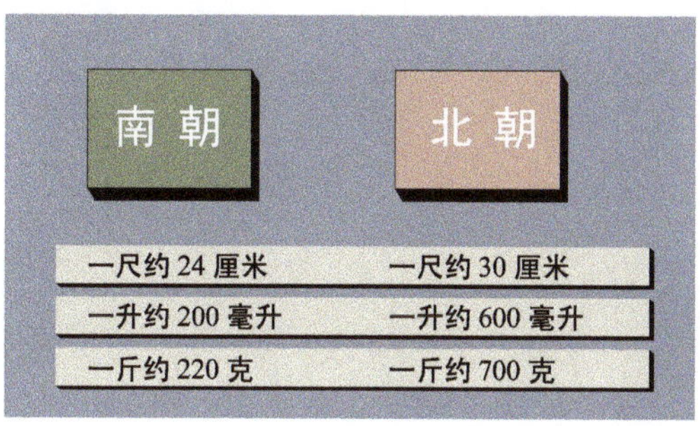

南朝与北朝度量衡单位量值之比

Comparisons of the values of units in the measuring and weighing systems between the Northern Dynasty and the Southern Dynasty

约 440 克。今藏河北省博物馆两件北魏时铁权，分别重 515.5 克和 593 克，据考证均系北魏时的一斤权。至隋统一度量衡时，每斤已高达“以古秤三斤为一斤”，推算每斤约 660 克。

由于南朝与北朝度量衡单位量值有数倍之差，故出现了“南人适北，视升为斗”的怪现象。南北朝度量衡的混乱可见一斑。

南北朝度量衡单位量值增长不平衡，究其原因，试作如下探讨。东晋以后，以长江为界，分为南朝和北朝。南朝仍为汉族人掌握政权，朝廷律历的制定以及政治、经济若干措施大抵沿袭汉、晋以来的旧制。度量衡制也基本保持不变，单位量值仅有缓慢增长。而北朝的北魏由拓跋氏建立。他们原居于漠北之荒野，以畜牧和射猎为业，当时的社会结构仍处在原始社会后期，仅凭着残酷屠杀和掳掠，统一了黄河流域，建立起北魏。建国后的 100 多年中，并未着力建制立法来管理国家。官吏没有俸禄，初期仅靠战争掠夺来的财富分赏官兵。建国后官吏贪赃枉法，自行搜刮勒索，政治极为腐败。度量衡更没有严明的制度，管理尤为混乱。为了多取于民，官吏与地主勾结，任意制造长尺、大斗、重秤而不受制度约束，从而造成度量衡单位量值成倍增长。待朝廷发现问题之严重而下发诏书统一度量衡时，却已无回天之力了。继北魏之后的东魏、西魏、北齐、北周皆历时甚短，混乱程度有过之而无不及。

南朝度量衡继承汉制，而北朝则因制度不健全，度量衡单位量值迅猛增长。故民间流传“南人适北，视升为斗”之说

In the Southern Dynasty,the measuring and weighing systems were the ones inherited from those in the Han Dynasty. In the Northern Dynasty, however, owing to its imperfect systems, the unit values of measures and weights increased dramatically. Hence there was a saying circulating among the people that “people in the South,when adopting the systems changing in the North, had to take sheng as dou”

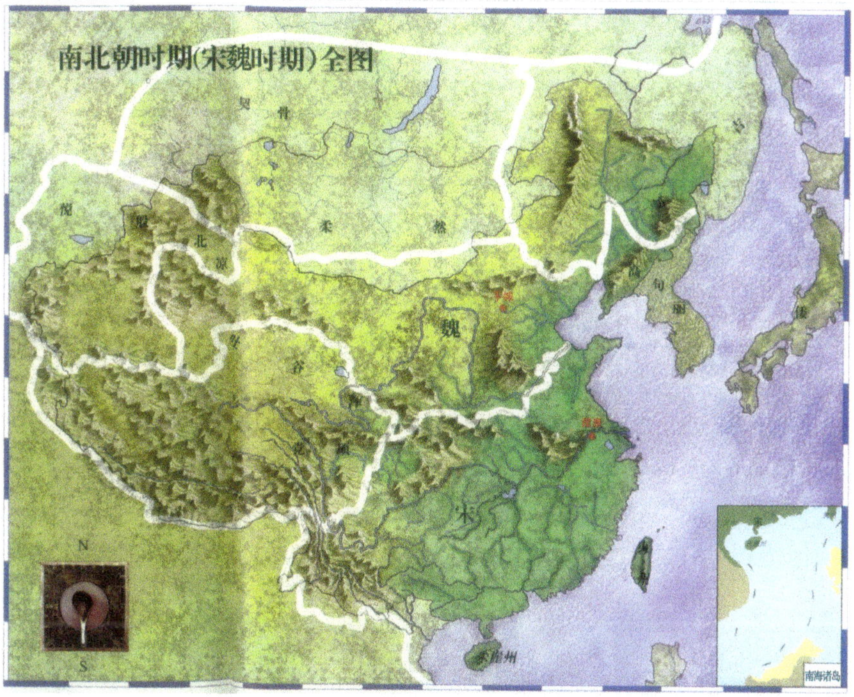

南北朝时期(宋魏时期)全图
A map of the Northern and Southern Dynasties (Song and Wei times)

2. The Measuring and Weighing System in the Northern and Southern Dynasties

The Northern and Southern Dynasties(A.D. 420-589,南北朝) comprised Song(宋), Qi(齐), Liang(梁) and Chen(陈) of the Southern Dynasty (南朝), and Northern Wei(北魏), Eastern Wei(东魏), Western Wei (西魏), Northern Qi(北齐) and Northern Zhou(北周) of the Northern Dynasty (北朝). In general, there were two characteristics in the measuring and weighing system in the Northern and Southern Dynasties (南北朝). First, values of units increased drastically . Second, the systems were in a great confusion. Take the linear measurement, for example. In the Tang Dynasty (唐代), an astronomer and mathematician named Li Chunfeng (李淳风) once categorized chi (尺) in different dynasties into 15 grades (十五等尺) in accordance with their different values. Historians of different times made a lot of textual researches on his classifications. Based on his findings, it can be calculated that the chi (尺) in Song (A.D. 420-479,宋) of the Southern Dynasty was 24.5 cm, which was almost the value of the chi (尺) in the Qin and Han dynasties (秦汉) but was 2% longer than the value in the Eastern Jin (东晋). This Song chi (宋尺) was adopted in Qi, Liang and Chen (齐, 梁, 陈). But in the Northern Dynasty (北朝) the value of chi (尺) was quickly increasing, thus causing tremendous confusion. For example, the value of chi (尺) was 25.5 cm in the initial period of Northern Wei (A.D.386-534,北魏), but turned to 27.9 cm in the intermediate period of the dynasty, and increased to 29.5cm towards the last years. By the time of Eastern Wei (A.D. 534-550,东魏), the value even increased to 30 cm. Northern Qi (A.D.550-577,北齐) inherited institutions from Eastern Wei after it overthrew the regime. In summary, throughout the 100 years of the Northern Dynasty (北朝), the value of chi (尺) increased by 6cm.

The situation of the measurement of capacity and weight, in terms of the increase in rate, was similar with that of length. In Liang (梁) and Chen (陈), the value of sheng (升) was the same as that in the Xin Mang (新莽) times, which was equal to 200 ml. In Qi (齐), 1 sheng (升) was about 300

北朝·魏 铜砣 底部錾刻一"平"字
A bronze sliding weight in Wei of the Northern Dynasty engraved with a character "Ping" at the bottom

ml, which was a little greater than the value of 1 sheng (升) of ancient times. In the Northern Dynasty (北朝), the value of sheng (升) had increased rapidly. In Northern Wei (北魏), 1 sheng (升) was 400 ml, but when Sui Wen Di (隋文帝) unified China and established the Sui Dynasty (隋朝) it increased to 600 ml.

As for the weighing system, in Liang (梁) and Chen (陈) of the Southern Dynasty, 1 jin (斤) was equal to 220g. In Qi (齐) 1 jin (斤) was 330g. In the Northern Dynasty the value of jin (斤) increased much faster. According to historical materials available, the value of dou (斗) in Wei (魏) and Qi (齐) doubled what it was in the ancient system. Today in the Museum of Hebei Province there are two pieces of iron weights (铁权) made in the Northern Wei (北魏) times, which weighed 515.5g and 593g respectively. It was guessed that they were weights of 1 jin (斤). When China was again unified in the Sui Dynasty 1 jin (斤) was around 660g, according to our calculation.

Because the value of units in the Northern Dynasty (北朝) can be multiple of that in the Southern Dynasty (南朝), so it was recorded that "when adapting the situation in the North, a southerner would mistake the sheng (升) for dou (斗)". Such was just one aspect that revealed the confusion of the measuring and weighing systems in the Northern and Southern Dynasties at that time.

Clearly there existed tremendous differences in the measuring and weighing systems between the Northern Dynasty (北朝) and the Southern Dynasty (南朝). Reasons attributing to these differences are explored and analyzed as follows.The Southern Dynasty was governed by the Han ethnic group (汉族). Institutions were based on the old ones in the Han Dynasty (汉朝) and in the Western and Eastern Jin (西晋, 东晋) times, so was the measuring and weighing system. In the light of the value of units, there was only some moderate increase. Things changed when the Northern Wei (北

北朝 • 魏 铁砣 分别重 181 克、207 克、222 克
Iron sliding weights in Wei of the Northern Dynasty, 181 g, 207 g and 222 g in weight respectively

南朝 • 梁 凤纹铜矩尺(线描图) 一尺长 24.9 厘米
A bronze square ruler with phoenix-shaped grains in Liang of the Southern Dynasty. The length of 1 chi was 24.9 cm

北朝 • 魏 铜尺 长 30.9 厘米
A bronze ruler in Wei of the Northern Dynasty, 30.9 cm in length

魏) Dynasty was established by the Xianbei (鲜卑族) tribe. Throughout the period of this regime that lasted for more than one hundred years, rules and regulations for the management of the country were never stipulated. Officials of the regime got no salaries from the central government but lived by plunder in the earlier stage. In measures and weighs there were no strict rules or regulations and the management was in confusion. As the corruption grew to its extremity, in order to exploit people as much as possible, officials colluded with landlords to make longer rulers (长尺), bigger dou (大斗), and heavier weights (重秤) according to their wishes instead of laws, so the value of units doubled and redoubled. When the central government noticed the problems and ordered to standardize the measures and weights, it was already too late. In Eastern Wei (东魏), Western Wei (西魏), Northern Qi (北齐), and Northern Zhou (北周), with each lasting for only a short period of time, the confusion of the measures and weights was even more serious.

南朝・宋 铜尺 残长 12.4 厘米，一尺长 24.7 厘米
A bronze ruler in Song of the Southern Dynasty, 12.4 cm in length (incomplete). The length of 1 chi was 24.7 cm

南朝・梁 镂刻鸟兽纹铜尺（拓本）长 24.8 厘米
A bronze ruler (in rubbing forms) in Liang of the Southern Dynasty, 24.8 cm in length, engraved with grains of birds and animals

北朝・北齐 铜砣 重 74 克
A bronze sliding weight in Northern Qi of the Northern Dynasty, 74 g in weight

南朝・宋 骨尺 长 24.8 厘米
A bone ruler in Song of the Southern Dynasty, 24.8 cm in length

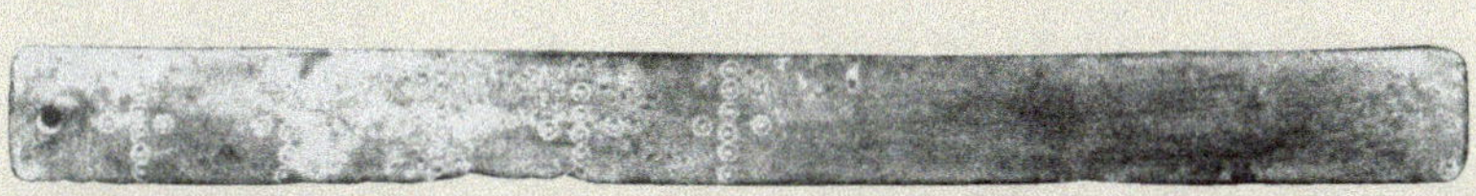

南朝 铜尺 长 25 厘米
A bronze ruler in the Southern Dynasty, 25 cm in length

3. 隋代统一度量衡和度量衡的大制、小制

581年，隋文帝杨坚建立了隋朝(581～618)，结束了割据纷争局面，中国重新统一。度量衡也随着政权的统一而再次统一起来，即将经过北朝已增大的度量衡用法律的形式固定了下来，结束了混乱无定制的局面。因此隋代在中国度量衡史上的地位也是十分重要的。

3. The Co-existence of the Two Systems in the Sui Dynasty

In the year 581, Yang Jian (杨坚) established a new dynasty named Sui (隋) that lasted from 581 to 618. From then on, Yang Jian was known as Emperor Wen Di (文帝). He ended the political division of China. China was unified again. In the history of Chinese measures and weights, the Sui Dynasty (隋代) was a landmark: the measuring and weighing system of the Northern Dynasty was fixed by laws, and the measures and weights in China were standardized again.

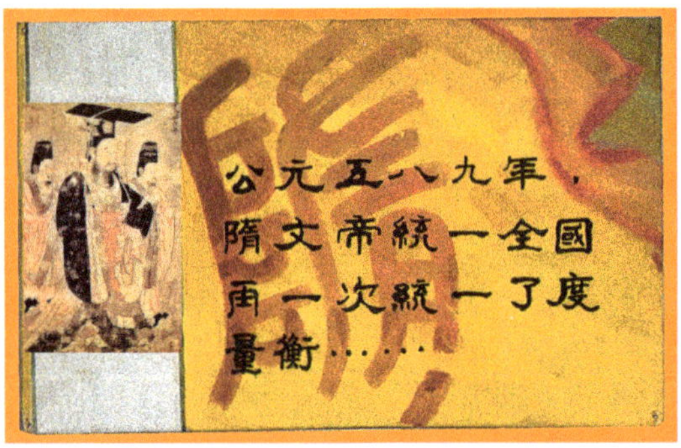

589年，隋文帝统一南北朝后，两次下令统一全国度量衡
After Sui Wei Di unified the Northern and Southern dynasties in A. D. 589, he issued edict twice to unify the measuring and weighing systems all over the country

隋代仁寿四年(604) 铜砣拓本
The bronze sliding weights made in the 4th year of Ren Shou (A.D. 604) of the Sui Dynasty, in rubbing forms

隋代 青釉兔纽瓷砣 重873克
A green-glazed china sliding weight, made in the Sui Dynasty, with a rabbit-shaped knob

(1)隋代统一度量衡

隋代的开国皇帝杨坚是一个生长在北朝并有一定权势的汉族官吏，因此对旧政权的腐败、制度的混乱都有深入了解。在他即位后，用七八年的时间，果断地进行政治、经济各方面的改革。面对复杂的社会状况，首先调整中央与地方机构，整顿吏制，修改刑律。为了恢复生产和繁荣经济，又下令统一货币、统一度量衡。据文献记载，隋初开皇年间(581～601)统一度量衡是以北朝最后一个王朝——北周(557～581)的制度为统一的标准，即1尺长约29.5厘米，1升约600毫升，1斤约660克。以此定为官府和民间日常用器的量值，史称度量衡之“大制”。

在中国历史上，秦始皇统一度量衡对秦以后的两汉、魏、晋影响深远。至隋文帝，又把经过南北朝而混乱的度量衡再一次统一起来，并且被隋以后强大的封建王朝唐、宋所沿袭，直至明、清都没有大的变化。

隋代 木尺 长29.7厘米
Wooden rulers in the Sui Dynasty, 29.7 cm in length

隋代 度量衡分大制和小制。尺度日常用大尺，长约29.5厘米，天文乐律则用小尺，长24.2厘米。此后被唐、宋、元、明、清历代沿袭
Throughout the Sui Dynasty, the measuring and weighing system comprised Da Zhi (Large System) and Xiao Zhi (Small System). Rulers for daily use were adopting the Large System, 29.5 cm in length; while rulers used in astronomy and temperament were adopting the Small System, 24.2 cm in length. The two systems were carried on in the following Tang, Song, Yuan, Ming and Qing dynasties

(1) The Standardization of Measures and Weights

Growing up in a family whose members were governmental officials in the North Dynasty, Yang Jian (杨坚) had full understanding of the situations where corruption and confusion were prevailing. Hence he decisively implemented innovation in the political and economic structures throughout his regime that lasted about eight years. Faced with complicated social situations, he started first to adjust the institutional structures in the central government and in the local ones. To boost the economy, he ordered to unify the currency and measuring and weighing systems. From historical works we find that the standardization of measures and weights was on the base of the system of Northern Zhou (A.D.557-581, 北周) which was the last regime of the Northern Dynasty in which 1 chi (尺) was 29.5cm, 1 sheng (升) was 600ml, and 1 jin (斤) was 660g. Historians referred to this system as the Da Zhi (大制 ,Large System).

In the Chinese history, the unification in the measuring and weighing system made by Qin Shi Huang (秦始皇) had far-reaching influences after the Qin Dynasty (秦代) into the Western and Eastern Han (西汉, 东汉) and Wei and Jin dynasties (魏朝, 晋朝). Emperor Wen Di (文帝) of the Sui Dynasty (隋朝) unified the measuring and weighing system again after the confusing period in the Northern and Southern Dynasties (南北朝). Just like the significance of what Qin Shi Huang had made, so the re-unification had great influences on the feudal dynasties that followed. The Sui system (隋制) was inherited by the Tang (唐), Song (宋), Ming (明), and Qing (清) dynasties.

隋代 权衡日常用大制，一斤约重 660 克；称药物、金银则用小制，一斤重约 220 克
In the Sui Dynasty, the Large System was adopted in daily weighing,with 1 jin equal to roughly 660 g; while the Small System was adopted for weighing medicine, silver and gold, with 1 jin equal to about 220 g

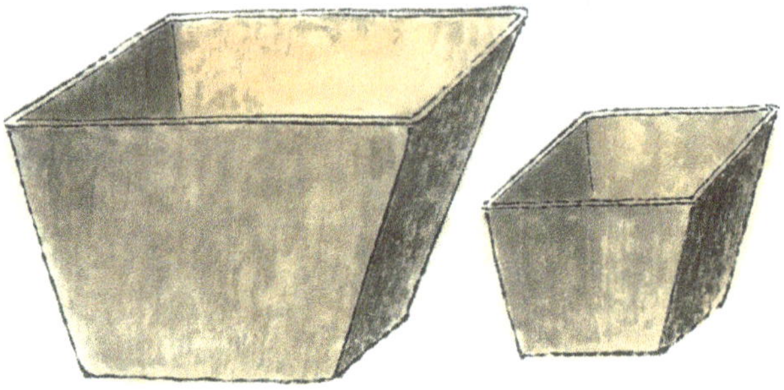

隋代 容量日常用大升，约容 600 毫升，小升约容 200 毫升
In the Sui Dynasty, the Large System was adopted in measuring capacity in terms of sheng, with 1 sheng equal to roughly 600 ml; while 1 sheng in the Small System was roughly 200 ml

北朝时期 敦煌壁画中的执秤图
Dunhuang frescoes of figures holding steelyards, painted in the Northern Dynasty

(2)度量衡大制和小制产生的原因以及量天尺的使用

大约在原始社会，先民们已用立竿测影的方法来判断时间的早晚。随着生产和生活的需要，人们创造了一种专门的测影工具——影(圭)表。它由一条南北水平的圭和置于圭的南端且与之垂直的表组成。圭面上刻有尺度，当太阳照射在表竿上，日影投在圭面上时，便可以根据圭面上的尺寸把影的长短记录下来，故圭表尺又有量天尺之称。《周髀算经》上说："夏至之影一尺六寸。"古人还常用正午表影校准漏刻，即"昼参诸日中之影"。反复测量并记录每次影在圭上的尺寸，就可以得知一年中日影最长的一天即是冬至，而最短的一天则是夏至。冬至、夏至确定之后，就可以得到回归年的长度，并可以预报季节、制定历法了。

东汉末至魏、晋，度量衡单位量值开始有所改变(长度从约 23.1 厘米增长至约 24.2 厘米，重量却从约 250 克下降至约 220 克)。中国古代遵奉着"天不变，道亦不变"的儒家信条，而观象授时是国之大政，量天之尺当然也不可轻易变动。秦汉时量天尺与民间日常用尺同长。东汉末至西晋初，民间用尺较古尺有所增长，每尺长约 24 厘米。中书监荀勖为考校古律必求古尺，得新莽时一尺的长度（约 23.1 厘米）。随后数学家祖冲之又考证了荀勖的律尺，认为确实合乎古制，从此测日影、定乐律都用古尺，但民间却用已增长的约 24 厘米的尺度。可见在西晋，量天尺与日常用尺已开始有了差别。东晋时战乱频

唐代 天文学家一行，主持天文大地测量，所用尺度系隋唐小尺。故小尺又有量天尺之称

Yi Xing, an astronomer in the Tang Dynasty, was in charge of deodetic and celestial survey. The rulers he used were adopted from the Small System prevailing in the Sui and Tang dynasties. Hence such rulers of the Small System were also called Rulers Measuring Heavens (Liang Tian Chi)

仍，礼乐制度混乱，度量衡标准器也失传。南朝元嘉年间(424 ～ 453)宋文帝诏太史令钱乐之重造浑天仪，但求古尺不得，便以当时民间尺为标准(长24.5 厘米)。于是量天尺与日常用尺的长度在南朝又取得了一致。与此同时地处北方的北朝，度、量、衡三者的量值仍在继续增长，1 尺已接近 30 厘米。隋文帝统一度量衡时，下令用南朝的小尺测日影，而用北朝的大尺作为官民日常用尺，形成了大、小两制。继隋之后的唐朝，把这既定的事实载入了《唐六典》。从此，唐代一行(张遂)测量子午线，宋代司天监测日影所用的圭表尺，元代科学家郭守敬造观星台所标的量天尺，都采用了隋、唐小制。今从明代制造的铜圭表上发现了当时量天尺的刻度，得 1 尺长 24.5 厘米，与钱乐之浑天仪尺度正好相符。量天尺在古代测日影、定历法中起到了重要的作用。在 1300 多年中，单位量值保持不变，从而保证了天文测量数据的连续性和稳定性。

南朝·梁 张僧繇绘执秤图
A copy of drawings of a figure holding a steelyard by Zhang Sengyao in Liang of the Southern Dynasty

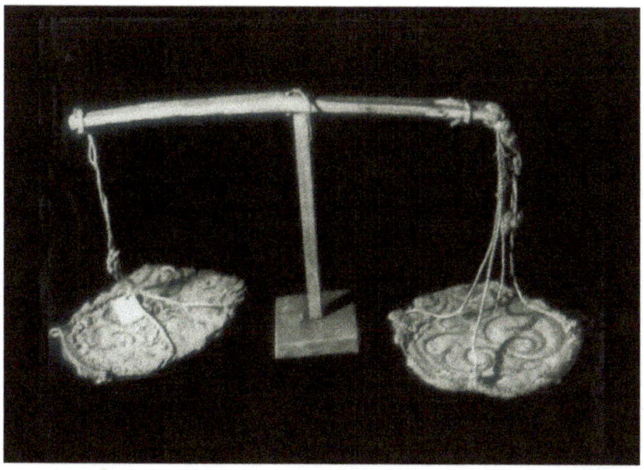

唐代 布天平
A scale in the Tang Dynasty

(2) Da Zhi (Large System) and Xiao Zhi (Small System) in the Sui Dynasty and Liang Tian Chi

In the pre-historic times, people told the time by observing the shadows of rods erected on the ground (立竿测影). As needs ascertained in their daily time, people then invented a time instrument called Gui Biao (圭表) which consisted of a stone compass and two sticks on it, one at the southern edge and the other at the northern edge. There were scales on the compass, and under the sun, the time was recorded as the shadows of the rods changed. Hence Gui Biao Chi was also known as Liang Tian Chi (量天尺), a primitive instrument for measuring time with the principle of using a Sundial (日晷). The other function of Liang Tian Chi was that it helped people to observe seasonal changes. In *Zhou Bi Suan Jing* (《周髀算经》), it was recorded that "it is Xia Zhi (夏至, Summer Solstice) when the length of the shadow is 1 chi (尺) and 6 cun (寸), the shortest and it is Dong Zhi (冬至,Winter Solstice) when the length of the shadow is the longest". By observing the changes of shadows and setting the seasons, people were able to calculate the solstice and foretell the changes of seasons in order to establish the calendar.

As mentioned earlier, the values of measures and weights changed during the last period of the Eastern Han (东汉) and during the times of the Wei and Jin dynasties (魏朝, 晋朝), where the unit length increased from 23.1 cm to 24.2 cm and weights decreased from about 250 g to about 220 g. As we are aware, from the ancient times, Chinese people observed and followed the Confucian beliefs that "while the providence (天) is unchanged, Dao(道) remains the same". Because meteo rology and time were crucial aspects in the governmental mandate, it was next to impossible to make any changes on Liang Tian

北魏时期至唐代 敦煌壁画中的天平
A balance in Dunhuang frescoes painted from the times of Northern Wei to the Tang Dynasty

Chi (量天尺).

In the Qin and Han dynasties (秦朝, 汉朝), unit value of Liang Tian Chi (量天尺) was the same as that of civil rulers. But during the last period of the Eastern Han (东汉) to the beginning of the Western Jin (西晋), unit value of civil rulers increased to around 24cm. In the Western Jin, Xun Xu (荀勖) made a textual research on ancient chi (尺) for the purpose of determining the exact value so as to calibrate temperaments. According to his research findings, the value of chi (尺) in the Xin Mang (新莽) times was around 23.1 cm. Afterwards a famous mathematician named Zu Chongzhi (祖冲之) examined the value of chi (尺) determined by Xun Xu (荀勖) and found that this value was identical with the value of the ancient chi (尺). With this verification, the value of the ancient chi (尺) was adopted in calculating time and in determining temperaments. Meanwhile, the increased chi (尺) with a value of 24cm was commonly accepted in civil linear measurement. This shows clearly that there existed differences in the values of lengths between Liang Tian Chi (量天尺) and civil rulers. In the Eastern Jin (东晋) the measuring and weighing system was in confusion for reasons such as frequent wars and customs in rites and temperaments. The worst thing was that the standards of values in the measuring and weighing system were lost in wars. During the Yuan Jia (元嘉) years, from 424 to 453 of the Song (宋) regime in the Southern Dynasty (南朝), Emperor Wen Di (文帝) decreed an official named Qian Lezhi (钱乐之) to rebuild the Hun Tian Yi (浑天仪, a Celestial Globe) based on the original shape. Since the value of ancient chi (尺) was unknown, the civil ruler with the value of length equal to 24.5 cm was applied in the measurement for construction. So in the Southern Dynasty (南朝), the values of length became identical again between Liang Tian Chi (量天尺) and civil rulers, which meant that the ancient chi (尺) was formally abolished. As a contrast, in the Northern Dynasty (北朝) the values of units in measures and weights kept increasing, and the value of the basic linear unit, chi (尺), was nearly 30 cm in the end. It is imaginable how confusing and chaotic the situations were at that time. During the time when Emperor Wen Di (文帝) of the Sui Dynasty (隋朝) was on the throne, efforts were made on standardizing

图间插文 Descriptions matching the Pictures

目前所见古代有关杆秤的绘画，时代最早的是南朝(梁)画家张僧繇的执秤图，敦煌壁画中的执秤图。其后有宋代大型杆秤称盐图，元代壁画中的执秤官称鱼图等。元代秤砣的大量出土，说明杆秤使用更为普遍并有了定量秤砣。

The earliest drawings of steelyards used in ancient China ever seen were: those drawn by Zhang Sengyao, an artist in Liang of the Southern Dynasty; the ones among the Dun Huang frescoes of figures holding steelyards; the ones in the Song Dynasty of steelyards in large scale weighing salt; the ones among wall paintings in the Yuan Dynasty of officials holding steelyards to weigh fish. That sliding weights in the Yuan Dynasty were unearthed in large quantity shows that steelyards were becoming more and more popular and sliding weights were produced with fixed values of weights.

宋代《政和政类本草》中的大型杆秤称盐图

A diagram of steelyard weighing salt in a large scale in a book entitled *Zheng He Zheng Lei Ben Cao* in the Song Dynasty

the measuring and weighing system. Imperial edicts were made known that Xiao Zhi (小制, Small System) of the Southern Dynasty (南朝) had to be used in time reckoning; and Da Zhi (大制, Large System) of the Northern Dynasty (北朝) had to be used in the linear measurement for civil purposes. Therefore during the Sui Dynasty the two systems were co-existing. In the Tang Dynasty (唐朝) the co-existing systems established in the previous dynasty was inherited and recorded in *Tang Liu Dian* (《唐六典》, *The Six Decrees and Regulations of the Tang Dynasty*). Since then linear measuring apparatus were all based on Xiao Zhi (小制, Small System), such as the one used by Zhang Sui (张遂) of the Tang Dynasty in measuring meridians; Gui Biao Chi (圭表尺 a ruler used in measuring the length of the shadow on the Gui Biao) of the Song Dynasty (宋朝); and Liang Tian Chi (量天尺) used by a renowned scientist named Guo Shoujing (郭守敬) of the Yuan Dynasty (元朝) in building the observatory.One of the bronze sundials made in the Yuan Dynasty (元朝) bears scales of Liang Tian Chi (量天尺). So the value of the chi (尺) in Xiao Zhi (小制, Small System) at that time was calculated as 24.5 cm, which was in conformity with the value of the chi (尺) in Qian Lezhi's (钱乐之) Hun Tian Yi (浑天仪, Celestial Globe). Liang Tian Chi (量天尺) played an important role in checking time on Gui Biao (圭表, Sundial) and in the promulgation of calendar. The unit values remained stable for a long period of time as 1,300 years, which guaranteed the continuity and stability of the astronomical observation in ancient China.

明代 炼丹图 调配各种矿物原料的比例，是炼丹术的关键工艺。此系清绘本《本草品汇精要》中的秤秤图

A diagram of getting extraction in pellet form prepared with heat, in the Ming Dynasty. Setting allocations of raw minerals was the key technique. This copy of drawing was seen in a book entitled *Ben Cao Pin Hui Jing Yao,* with a diagram of a steelyard

元代 山西广胜寺壁画 执秤官称鱼图

Paintngs of officials holding steelyards to weigh fish among the wall paintings in Guang Sheng Temple, Shanxi Province, in the Yuan Dynasty

THE HISTORY  OF ANCIENT CHINES

THE HISTORY OF ANCIENT CHINESE MEASURES AND WEIGHTS

中国古代计量史

IEASURES AND WEIGHTS

八、唐宋时期的度量衡

Chapter Eight

Measures and Weights in the Tang and Song Dynasties

1. 唐代的度量衡

建立于公元 618 年的唐朝，在中国历史上占有显赫的地位。唐朝在中国度量衡发展史上，除了在严格的管理制度、单位制的改革上占有重要地位外，对东方各国的影响也十分深远。

唐朝的典籍《唐六典》中，对度量衡制度有比较详细的记述。大意是以汉代度量衡为典范，以累黍求得度量衡之标准。即横累 100 黍为一尺之长，1200 黍的容积为一龠之量，100 黍之重为一铢。仍沿袭汉代单位制为唐制。但是从东汉末至唐初已经过了近 400 年的岁月沧桑，度量衡又经历了北朝的混乱和隋代的再次统一，百姓已习惯使用隋制，单位量值不可能再回复到秦汉时期。《唐六典》便根据这一具体情况把从西晋时逐渐形成，至隋代已记录在案的大、小制用法典的形式再次规定下来。即古制是典范、是源泉，而经隋代统一后的度量衡则为大制。大制与小制之间有明确的比例，即 1 尺：1.2 尺，1 升：3 升，1 斤：3 斤。小制用来测量圭表的影长以定历法，确定黄钟律管之长以定雅乐以及考定古代礼制(如车、服、冠等)和称量药物。除此之外，官民均用大制。《唐六典》的记载是我们了解唐制的重要依据。

今所搜集到的唐尺已达 30 余支，其长度在 29～31.8 厘米。由于尺是日常用器，唐代历时近 300 年，遗存至今的各种尺，使用的时间、地点以及当时对精度的要求等各种条件今天均无法确定，因此这些尺之间存在一定的误差也就在所难免了。目前我们根据所见各种文献资料，更着重对唐尺本身的综合分析研究，将一尺之长厘定为 30 厘米。

唐尺所见甚多，而容量迄今却未见任何比《唐六典》更确切的资料和实物可供进一步考证。根据《唐六典》："以古斗三斗为一斗"的粗

唐 拨镂白牙尺 长 30 厘米
An ivory ruler with engravings in the Tang Dynasty, 30 cm in length

唐代 拨镂红牙尺 长 29.7 厘米、29.5 厘米
Red ivory rulers with engravings in the Tang Dynasty, 29.7 cm and 29.5 cm in length respectively

略记述，从秦汉至魏晋 1 升均合 200 毫升的标准去推算，厘定唐一升约合 600 毫升。

迄今也未见唐代砝码流存至今。幸得近年来考古发掘出土一批有自重铭刻的金银器可供考证。如陕西省出土的一批税银，分别刻有 5 两、10 两、20 两、50 两等。此外还有一些有记重铭刻的日用金银器。经过实测和对各类数据的合理筛选，得 1 斤平均值在 660 克左右，与《唐六典》所说“以古三斤为一斤”相合。据考证，东汉末，重量单位值略有下降，1 斤约合 220 克，魏晋及南北朝沿用。

唐代长度、容量单位制均沿袭前朝旧制，唯重量单位有所改进。秦汉以来，长度和容量皆为十进制，而重量不是十进制。唐初铸“开元通宝”钱，表示开辟新纪元及通行宝货之意。《唐会要》又把“开元通宝”的直径和重量做了明确规定：“径八分，重二铢四累”。二铢四累正合一两的十分之一，即 10 枚重 1 两。此后，人们感到 10 钱为一两比二铢四累为一两更便于计算，于是约定俗成地出现了一个新的单位“钱”。宋代又将十进位的分、厘取代了钱以下的累、黍。这样重量单位除了仍用 16 两为一斤，120 斤为 1 石之外，两以下的其他单位都采用了十进位。

唐代对度量衡有严格的管理制度。如《唐会要》中规定了主管、制造、颁发以及检定的机构。除了行政措施外，政府还将度量衡管理条文载入法律文书《唐律疏议》之中，制定了严格的法律条文和奖惩办法，如：凡营私舞弊者，皆严惩不贷。凡私造度量衡器具并贪污国库资财者，皆受各种笞刑直至处死。这也是所见中国历史上将度量衡法令载入国家法律条文之首例。

唐代 鎏金花卉铜尺 残长 24 厘米，一尺长 30.8 厘米
A gold-gilded bronze ruler carved with flowers and plants, 24 cm in length (imcomplete) in the Tang Dynasty. 1 chi was equal to about 30.8 cm

唐代 拨镂绿牙尺 长 29.5 厘米
A green ivory ruler with engravings in the Tang Dynasty, 29.5 cm in length

1. The Measuring and Weighing System in the Tang Dynasty

The Tang Dynasty (唐朝) was established in the year 618. Its broad territory, prosperous economy and colorful culture made it an eminent dynasty in the Chinese history. The Tang Dynasty took an important position in the development of measures and weights in China for its strict management institutions and reformations in the unit systems, as well as its profound influence exerted on oriental countries.

According to the records in *Tang Liu Dian* (《唐六典》), the Tang Dynasty followed the Han Dynasty as far as the system of measures and weights was concerned. There were, of course, some modifications. In the system of the Tang Dynasty, the standards were also fixed with Lei Shu (累黍): the length of 100 pieces of corn millet put side by side was 1 chi (尺); the volume of 1,200 pieces was 1 yue (龠); the weight of 100 pieces was 1 zhu (铢). All of these standards were the same as that in the Han Dynasty in terms of their values. Meanwhile, since people in the Tang Dynasty were accustomed to the system of the Sui Dynasty (隋朝), so the coexisting two systems Da Zhi (大制, Large System) and Xiao Zhi (小制, Small System), as mentioned earlier, in the Western Jin (西晋) times and in the Sui Dynasty (隋朝), were fixed in the royal codes. It was specified in the Codes that the ancient system, namely, Xiao Zhi (小制) was the sources and references for clarifications. The ratios between Da Zhi and Xiao Zhi were set as follows:

Da Zhi (大制) : Xiao Zhi (小制),1 chi (尺) : 1.2 chi (尺), 1 sheng (升) : 3 sheng (升), 1 jin (斤) : 3 jin (斤).

It was stipulated that Xiao Zhi was used for measuring the shadows on Gui Biao (圭表) and for setting the calendar accordingly and for cross-checking the length of the Huangzhong Pitch Pipe (黄钟律管) in order to determine the royal music and rites processes. It was meant for weighing herbs in medicine as well. Other than that, Da Zhi (大制) should be implemented both for governmental officials

唐代 鎏银花卉铜尺 长 31 厘米
A gold-gilded bronze ruler carved with flowers and plants, 31 cm in length, in the Tang Dynasty

唐代 龙纹铜尺 长 29.7 厘米
A bronze ruler carved with grains of dragons, 29.7 cm in length, in the Tang Dynasty

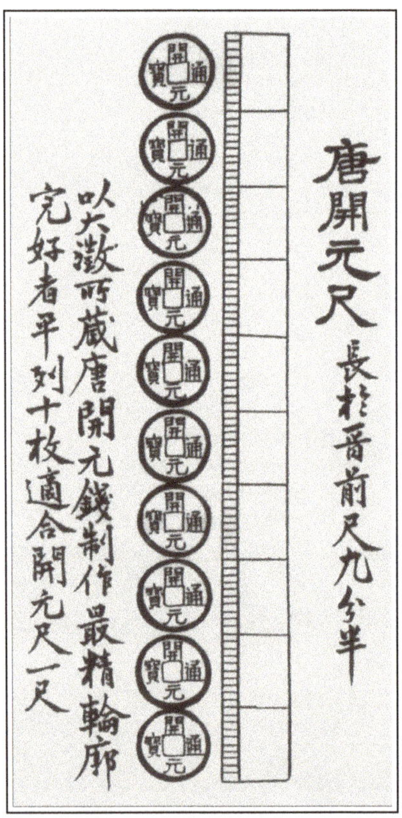

清代 吴大澄绘以 10 枚开元通宝排列长 25 厘米,考为唐小尺
A painting by Wu Dacheng, in the Qing Dynasty, of 10 coins aligned with a length of 25 cm, which was examined as of chi in Small System of the Tang Dynasty

唐代 鎏金舞马壶 铭刻“重十三两半”
A gold-gilded kettle imprinted with lively horses, with inscriptions of “13 and Half Liang in Weight” in the Tang Dynasty

唐代 刻花金碗 铭刻“九两半”
A gold-gilded bowl imprinted with flowers, with inscriptions of “9 and Half Liang” in the Tang Dynasty

and common people alike. *Tang Liu Dian* (《唐六典》) provides us with reliable documents for studying the systems in the Tang Dynasty.

Measuring the lengths of the 30 rulers of Tang available at present, we have observed that they varied between 29 cm and 31.8 cm. Because rulers were meant for daily use, and the Tang Dynasty lasted for about 300 years, it is hard to determine the actual years when, and places where they were used. Also, it is difficult to find out the requirements on precision at different times throughout the Tang times. Therefore, the variations were inevitable. Judging by the results of our comprehensive study and analyses on all kinds of historic documents available, we found it quite reasonable to claim that the value of length for 1 chi (尺) is confirmed as 30 cm.

With regards to the capacity of sheng (升), according to what was recorded in *Tang Liu Dian* (《唐六典》), “taking 3 ancient dou (古斗) as 1 dou (斗)”; to follow suit, the capacity of 1 sheng (升) was equal to 200ml in the ancient times, hence the capacity of 1 sheng (升) in the times of Tang was approximately 600 ml.

As for the units of weighing, there has been no objects ever found that were made in the Tang Dynasty. Luckily we got some references from what was unearthed in recent years, a series of golden and silver objects. For example, a bunch of silver for tax levying was unearthed in Shaanxi (陕西) Province with various numbers in liang (两) carved on them. The numbers read: 5 liang (两), 10 liang (两), 20 liang (两)

唐代 铜砣
A bronze sliding weight in the Tang Dynasty

and 50 liang (两), etc. There were also certain numbers of golden and silver objects for daily use, with units of weights on them. Based on our selections of various kinds of data, we reached the average value of jin (斤) as 660 g. According to our findings with regard to units of weights prevailing during the last period of the Eastern Han (东汉), the value of weights declined then, resulting in 1 jin (斤) equal to 220g. Such a value was recognized and implemented throughout the Wei and Jin dynasties (魏, 晋), and the Northern and Southern Dynasties (南北朝), when the value of 1 jin (斤) was equal to 660g. This was in conformity with what was stated in *Tang Liu Dian* (《唐六典》), "taking 3 ancient jin (斤) as 1 jin (斤)".

In the Tang Dynasties (唐朝), the units of length and capacity were the same as those in the earlier dynasties, but the units of weights were changed. Since the Qin and Han dynasties (秦, 汉), the decimal system had been adopted in expressing the units of length and capacity, except the units of weights. In the early period of Tang, the typical characters carved on the coins were Kai Yuan Tong Bao (开元通宝), meaning the new currency indicated a new era with treasures flowing in. It was clearly stipulated in *Tang Hui Yao* (《唐会要》) that the diameter and weight of the coin with Kai Yuan Tong Bao was set at 8 fen (分), and 2 zhu (铢) & 4 lei (累) respectively. The weight of 2 zhu (铢) and 4 lei (累) was exactly 1/10 of 1 liang (两), that is, the total of 10 coins weighed 1 liang (两). Later on, people felt that it would be easier to calculate 10 qian (钱) as 1 liang (两) than to calculate 2 zhu (铢) & 4 lei (累) as 1 liang (两), hence a new unit of qian (钱) came into being. In the Song Dynasty (宋), the units of lei (累) and shu

唐代 银盘 墨书"十四两三分"
A silver dish with characters in China ink as "14 Liang and 3 Fen" in the Tang Dynasty

唐代 银碗 墨书"合重三十六两,光明沙二十一两"
A silver bowl with characters in China ink as "36 Liang in Weight, including 21 Liang of Shining Sand" in the Tang Dynasty

唐代 赋税银饼 铸"庸调银十两"
A cake-shaped container for taxation in the Tang Dynasty, cast with "10 Liang of Yong Tiao Silver"

图间插文 Descriptions matching the Pictures

通过对有自重铭刻器物的实测,综合分析得唐代1斤重约660克。
After examining such containers with carved inscriptions of their weights, and collating the values of the units, we found that 1 jin in the Tang Dynasty was about 660 g.

(黍) with lesser values than qian (钱) were replaced with the decimalized fen (分) and li (厘). From then on, except for the weights units of 1 jin (斤) equal to 16 liang (两) and 1 dan (石) equal to 120 jin (斤), all the other units were decimalized.

The Administration in the Tang Dynasty adopted a strict management process in controlling the measuring and weighing system. As was stated in *Tang Hui Yao* (《唐会要》), the authoritative organizations for producing, distributing and inspecting the system were clearly specified. Apart from administrative methods, the then central government set the measuring and weighing system in legal items in the collections of law documents *Tang Lv Shu Yi* (《唐律疏议》), with specifications on rewarding and penalizing methods. This was the first in the Chinese history to set the measuring and weighing system in the governmental law documents.

唐代 银碗 墨书"十一两一分强"
A silver bowl in the Tang Dynasty with charaocters that read "11 Liang 1 Fen and a little more"

唐代 法律文书《唐律疏议》中明确规定有关度量衡的法律条文：凡使用不准确的度量衡器者，处以笞、杖等刑
In a legal document in the Tang Dynasty, *Tang Lv Shu Yi*, legal articles were clearly stipulated with regard to the implementation of the measuring and weighing system: those who use inaccurate apparatus for measuring and weighing would be penalized with caning or flogging

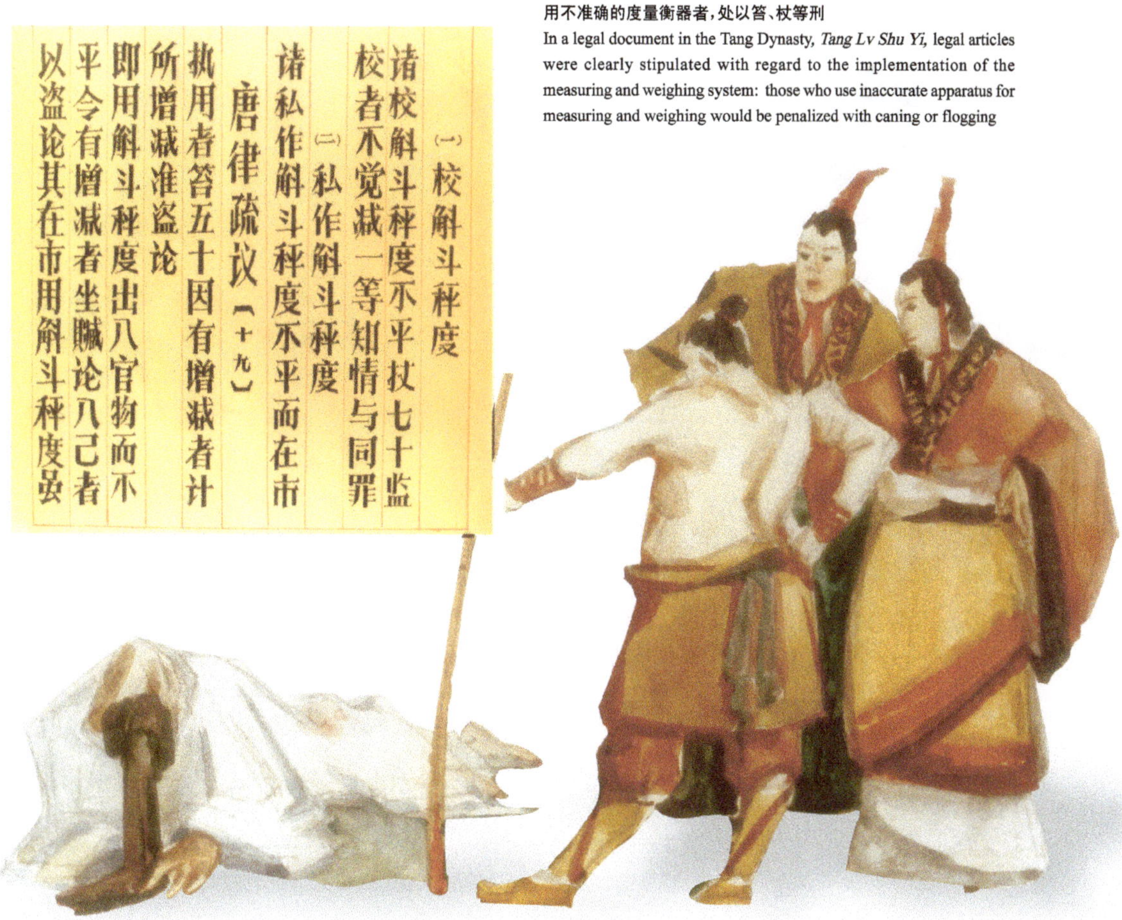

(一)校斛斗秤度
诸校斛斗秤度不平杖七十监
校者不觉减一等知情与同罪
(二)私作斛斗秤度
诸私作斛斗秤度不平而在市
唐律疏议〔十九〕
执用者笞五十因有增减者计
所增减准盗论
即用斛斗秤度出入官物而不
平令有增减者坐赃论入己者
以盗论其在市用斛斗秤度虽

2. 宋代的度量衡

唐代以后又出现了一个短暂而又混乱的五代十国(907 ～ 960)。其后赵匡胤(史称宋太祖) 建立了宋朝。宋朝分北宋(960 ～ 1127)和南宋(1127 ～ 1279)两个阶段,共 319 年。与宋朝并存的还有辽、金、夏。宋朝经济也曾有过比较繁荣兴盛的时期,农业产量提高,手工业、商业也呈现出一片繁荣景象,瓷器远销海外,科学技术仍居世界领先地位。中国航海史上首次使用先进导航仪器——指南罗盘,它对世界航海技术的进步起到了革命性的影响。这一时期的度量衡既在某些方面比前朝有所改进,又在一定程度上因管理不力而相当混乱。

北宋时期 店下样石质权 刻铭文 38 行 293 字,详述此权是为了防止盐务运输过程中发生偷盗而设置于鬻盐邸店的专用砝码。权重 140 千克,相当于宋衡制 220 斤

A Dian Xia Yang stone weight in the Northern Song Dynasty, carved with 293 characters in 38 lines, stating in details that such weights were used for weighing salt as a preemptive method in case the salt was stolen at the strations on the way of transportation. The weight was 140 kg in weight, about 220 jin in the weighing system in the Song times

北宋时期 熙宁(1068 ～ 1077)铜砣 刻铭文 15 行 167 字,详述了由贵池县按"官样"铸造的"壹佰斤"砝码。重 62.5 千克

A bronze weight made in the years of Xi Ning of the Northern Song Dynasty (A.D. 1068-1077), carved with 167 characters in15 lines, stating in details that it was a 100-jin weight (62.5 kg in weight), cast in Gui Chi County, following the suit of the official sample

2. The Measuring and Weighing System in the Song Dynasty

After the Tang Dynasty, China witnessed a period of chaos that was known as the Five Dynasties and Ten Kingdoms (五代十国, A.D.907-960) that lasted for a short period of time. In the year 960, Zhao Kuangyin (赵匡胤), the Emperor Tai Zu (太祖) of the Northern Song Dynasty (北宋) established the Song Dynasty (宋朝), which ruled the Northern Song Dynasty (A.D. 960-1127) and the Southern Song Dynasty (南宋, A.D. 1127-1279). During the 319 years of this regime, the Song Dynasty coexisted with another three regimes: Liao (辽), Jin (金) and Xia (夏).

Agriculture and handicraft industries made further progress during the Northern Song period. The yield of crops increased, the Song porcelain enjoyed a high reputation abroad, and the science and technology were taking a cutting edge in the world. It was in the Song Dynasty that the invention and usage of compass took place, and China was the first country in the world to use compass to guide directions for ships in navigation, which had revolutionary influence on the progress in navigating technology in the world. There were also some improvements in the measuring and weighing system in the Song Dynasty although some confusion occurred occasionally owing to mismanagement.

北宋时期　太宗(976 ~ 997 在位)下令统一度量衡。刘承珪（掌管国库收支的官吏）创制了两支小型戥秤，用作称量国库银两的标准衡器，并校准和颁行了成套标准砝码

Tai Zong, Emperor of the Northern Song (reigned A.D. 976-997) issued an edict to unify the measuring and weighing systems. Liu Chenggui (the official in charge of the national treasury) invented 2 small weighing apparatus called Dengzi as the standard instruments for weighing in managing the account of treasury. He also calibrated and enforced a set of standard weights

(1)北宋统一度量衡及刘承珪创制戥秤

北宋初创，宋太祖在把主要精力用于掌握兵权的同时，也注意到统一全国度量衡。公元976年太宗即位，再次下诏书统一度量衡。当时掌管国库的官吏刘承珪担起了此项任务。他首先检验了国库中所收藏的各种标准器，发现多有失准，造成使用中产生许多弊端和争讼。刘承珪经过反复研究后创制了一种小型精巧的杆秤，后世赋以专用名称——戥秤。为了保证其称量准确，刘承珪将戥秤分成两种规格，以便相互校准。一种采用秦汉古制，即两、铢、累、黍非十进位(10 黍 = 1 累，10 累 = 1 铢，24 铢 = 1 两)，最大秤量是 1 两，分度值为 1 累；另一种则采用自唐以来改革后的十进制的钱、分、厘、毫制，其最大秤量为一钱半，分度值为一厘。两种戥秤相互参校便可以得到两、钱、分、厘和两、铢、累、黍等单位量值。再用一两戥秤称淳化年代制造的铜钱，选取每枚重量合二铢四累(一钱)者，积2400 枚来确定 15 斤的标准。再选用称量分别为一钱半、一两的两种戥秤和最大秤量为 15 斤的标准秤，重新铸造了一批成套的砝码作为最高标准置于官府，并颁发到全国各地。刘承珪的戥秤比一般杆秤小巧且称量精确，深受民众欢迎，成为此后称量金银、药材等贵重物品的专用工具而沿用了近千年。

金代 铜砝码 一面刻"大定十五年(1175)造典字号",另一面刻"壹佰两"。重 3962.6 克
A bronze weight, 3,962.6 g in weight, in Jin of the Southern Song Dynasty with inscriptions carved on one side as "The 15th year of Da Ding (1175)", "Zao Dian Zi Hao"; and on the other side as "100 Liang"

(1) The Unification of the Measuring and Weighing System in the Northern Song Dynasty

During the early period of the Northern Song Dynasty, while Emperor Tai Zu (太祖) mainly engaged himself in controlling all military power in the central government, he also paid attention to the unification of the measuring and weighing system. In the year 976, Emperor Tai Zong (太宗) ascended the throne. He issued an edict ordering the standardization of measures and weights. The official who was in charge of the court treasury and fiscal affairs named Liu Chenggui (刘承珪) was assigned the task. Liu conducted serious examinations on various kinds of standard apparatus, and found most of the apparatus were not accurate enough. He then devoted himself to the research on iron yard and invented a new type of balance that was later called Dengzi (戥子) which was a small and delicate balance for weighing precious metal, medicine, etc.

In order to guarantee the accuracy, Liu Chenggui further designed two categories of Dengzi on two weighing systems respectively for the purpose of cross-checking. One category of Dengzi was based on the ancient system in the Qin and Han dynasties (秦, 汉) that was a mixture of decimal system and duodecimal system wherein liang (两), zhu (铢), lei (累) and shu (黍) were not in the decimal system—10 shu (黍) = 1 lei (累), 10 lei (累) = 1 zhu (铢), 24 zhu (铢) = 1 liang (两). On the beam of such kind of Dengzi, there were scales in lei (累), which amounted to 1 liang (两) maximum. The other kind of Dengzi was based on the decimal system that was established in the Tang Dynasty with the basic units of qian (钱), fen (分), li (厘) and hao (毫). On the beam of this kind of Dengzi,

南宋时期 铜砣 一面铭刻"阜昌",另一面铭刻"新法"。"阜昌"是南宋初受金控制的傀儡政权刘豫的年号(1130～1137)
A bronze sliding weight in the Southern Song Dynasty with inscriptions carved on one side as "Fu Chang", and on the other side as "Xin Fa (New Standard)". Fu Chang was the reign title of Emperor Liu Yu, a puppet regime (1130-1137) under the control of the Jin Dynasty during the early years of the Southern Song Dynasty

there were scales in li (厘), which amounted to 1.5 qian (钱) maximum. With these two categories of Dengzi for double checking and cross-checking, the values of one set of units were determined in comparison with those of the other set, namely,liang (两), qian (钱), fen (分), and li (厘) against liang (两), zhu (铢), lei (累), and shu (黍), etc. Dengzi was also used for standardizing weights. First, with Dengzi, the ratio of the ancient system to that established in the Tang Dynasty could be found and the value of units in the two systems could be calculated. Second, with Dengzi, the standard of 15 jin (斤) could be established as such: Taking a Dengzi based on the ancient weighing system to weigh coins minted in Chun Hua (淳化) times and selecting 2,400 coins with each weighing 2 zhu (铢) and 4 lei (累), i.e., 1 qian (钱), the total weight of these coins was 15 jin (斤). The two categories of Dengzi with the standard iron yard with the maximum scale of 15 jin (斤) enabled the government to cast weights in series, both for storing in the central government as the standards or for distributing to local governments as a means of standardizing weights.

Credited to Liu Chenggui, Dengzi was a very popular weighing device in the market, for it was small and exquisite and with high accuracy. From then on, as a specialized tool for weighing such valuables as precious metals and medicines, Dengzi has been used for centuries.

北宋时期 嘉祐铜则 正面铭刻"嘉祐元年(1056)丙申岁造",背面刻"铜则重壹佰斤黄字号",周身刻缠枝牡丹纹。重 64 千克

A Jia You Tong (bronze) Ze (standard weight) in the Northern Song Dynasty carved with inscriptions on one side as "Produced in the first year of Jia You (1056) Bing Shen"; and on the back as "This bronze standard is 100 jin in weight". Engraved with grains of peonies around, it is 64 kg in weight

(2)李照乐秤

北宋皇帝对古律、雅乐颇感兴趣。据《宋史》记载，仁宋(1023～1063)屡颁诏书访乐、议乐。一时间宫廷上下研究乐律尺者甚众，既有朝廷重臣，又有名人雅士。名目繁多的乐尺、乐斗、乐秤充斥朝廷内外。仁宗之后的历代君主皆延此风。学者各持己见，趋者良莠不齐。其中值得一提的是李照设计的乐秤。景祐二年(1035)李照因奉诏改乐而研制了乐律管、乐尺、乐量和乐秤。李照研制的乐秤，以水的相对密度为标准，故又称水秤。水秤以1合(10分之1升)水之重定作1两，1升水之重定作1斤。即利用水的物理特性来定单位重量。这种以自然物质定作重量标准的构想是科学的。此外，他还提出将16两为1斤改成10两为1斤。然而，囿于封建经济和因循守旧思想的束缚，李照这一创造性的改革建议，因受到众多非议而未能在实际中运用。

(2) Li Zhao's (李照) Yue Cheng (乐秤)

In the Northern Song (北宋), the emperors were very interested in the ancient temperament and royal music. It was recorded in *Song Shi* (《宋史》) that Emperor Ren Zong (仁宗, A.D. 1023-1063) had issued many edict orders to explore the ancient music. Then for a period of time some powerful officials and celebrities with no official posts were engaged themselves in the musicology and the study of temperament. As a result various kinds of pitch-originated steelyard (乐秤), pitch-originated chi (尺) and pitch-originated capacity-measuring apparatus emerged. Among them a kind of pitch-originated steelyard designed by Li Zhao (李照) merits attention. In the second year of Jing You (景祐, A.D.1035), Li Zhao received an edict to make a reform on the music of the time, so he designed pitch pipes and some pitch-originated measuring and weighing apparatus. In designing the pitch-originated steelyard, Li Zhao related volume of water to the weighing units. So his steelyard was also called water steelyard (水秤). In his unit system, the weight of 1 ge [合, equal to 0.1 sheng(升)] of water was defined as 1 liang (两) and the weight of 1 sheng (升) was 1 jin (斤). At that time Li Zhao's idea of taking a natural object as weighing standard was science based. Meanwhile, Li Zhao suggested to make 1 jin (斤) equal to 10 liang (两) instead of 16 liang (两). But traditional ideas weighed down his creative ideas and they were unable to be applied in social life.

宋代 铜砣
A bronze sliding weight in the Song Dynasty

金代 铁砣 一面铭刻“明”，另一面铭刻“昌”。“明昌”是金章宗完颜璟(1190～1208)的年号
An iron sliding weight in Jin of the Southern Song Dynasty carved with inscriptions on one side as “Ming” and on the other as “Chang”. “Ming Chang” was the reign title of Emperor Zhang Zong (Wan Yan Jing, 1190-1208) of the Jin Dynasty

辽代“嵩(崇)德宫”铜量 系宫廷用量器。器身三分之二处有隔档，上部容1047毫升，下部容500毫升
A Chong De Gong bronze instrument in Liao of the Northern Song Dynasty which was an apparatus of capacity used in the imperial court. There was a partition at the 2/3 level separating the upper part with a capacity of 1,047 ml from the lower part with a capacity of 500 ml

(3)两宋度量衡量值考

两宋时期的度量衡器，名目繁杂。名称多达数十种，单位量值也各不相同。但凡是由政府权威机构颁发的器具仍有统一的规定。地方上特殊用器，虽不与标准相符，却都有明确的比例关系。综合前人对尺度的研究，多在30.8～31.7厘米。今搜集到北宋尺的实物，长短虽也不齐，却也多在31～32厘米。今考定为31.4厘米。南宋官尺沿用北宋，唯地方特殊用尺出现长短不合规格的现象。

宋代的容器在形式上较前代有所改进，把原来上口大改为上狭底宽，使出入之间盈亏不会相差甚远。单位量值则略有增长，1升约为702毫升。

目前搜集到的两宋砝码及相关器物，经实测1斤在650克左右。而近年出土了北宋100斤铜则和铜砣各一件，1斤约合640克。经根据各种可信文献和实物综合分析，把两宋时期1斤单位值厘定为640克。

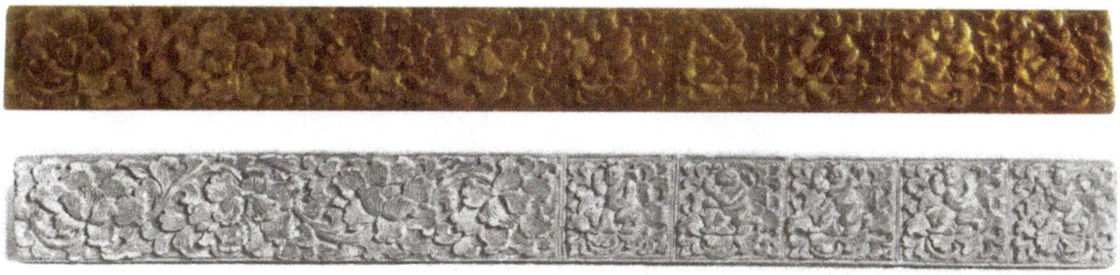

宋代 五子花卉木尺 长31.7厘米
A wooden ruler with grains of Wu Zi flowers and plants in the Song Dynasty, 31.7 cm in length

宋代 鎏金铜尺 长31.7厘米
A gold-gilded bronze ruler in the Song Dynasty, 31.7 cm in length

(3) Examination of the Weighing Units in the Song Dynasty

In the Song Dynasty (宋朝), measuring and weighing apparatus varied in names and magnitudes of units. Standard apparatus cast by the central government were unified in the magnitudes of units, but some special apparatus prevailing within a certain area were not identical with the standard ones in terms of the magnitudes of units, although there was well established unit system for each of these special measuring and weighing apparatus.Collecting and collating the results of researches conducted by former researchers, we found that the length of 1 chi (尺) had been mostly set in a range from 30.8 cm to 31.7 cm. Recently through measuring some unearthed rulers made in the Song Dynasty, we found that the length of 1 chi ranged from 31 cm to 32 cm. Recently, a synthetic study showed that 1 chi of the Song Dynasty was equal to 31.4 cm. In the Southern Song, the standard chi of the Northern Song (北宋) was adopted, except for certain local areas where differences existed in length among rulers for special purposes.

Capacity measuring apparatus in the Song Dynasty were different with those used in previous ages in shapes and magnitudes of units. Originally, a capacity measuring apparatus always had a wider top and a narrow bottom. It was renovated to a shape with a relatively narrow top and wider bottom. As for the magnitudes of units, 1 sheng (升) of the Song Dynasty was increased a little bit, to around 702 ml. The actual weighing of collected weights and related utensils made in the Song Dynasty showed that 1 jin (斤) of the Song Dynasty was equal to 650 g or so. But according to the actual weighing of recently unearthed bronze weights marked 100 jin (斤) cast in the North Song, the value of 1 jin (斤) was supposed to be 640 g. A comprehensive study based on historical works and actual objects shows that it is more reasonable to take 1 jin (斤) as 640 g.

THE HISTORY

OF ANCIENT CHINESE

EASURES AND WEIGHTS

THE HISTORY OF ANCIENT CHINESE MEASURES AND WEIGHTS

中国古代计量史

九、汉至唐宋时期计量测试技术的发展与进步

Chapter Nine

The Development of Measuring Technologies From the Han to the Tang and Song Dynasties

从汉至魏、晋、南北朝，出现了一批杰出的科学家，他们在数学、天文、计时和机械制造方面大量地使用计量技术，促进了科学技术的进步。

From the Han Dynasty (汉朝) to the Wei and Jin dynasties (魏朝, 晋朝), and the Northern and Southern Dynasties (南北朝), many outstanding scientists were emerging. They frequently applied the measuring technologies in mathematical and astronomical researches, in time calculating and in machinery making, which provided continual impetus to the development of sciences and technologies.

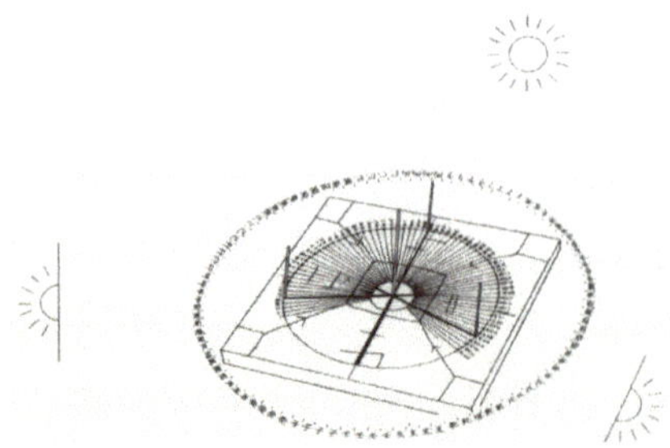

日晷的使用方法是，在石板中心圆孔处直插"正表"，外圆立"游表"，观测表影在辐射状标志线间的位置
The application of sundial: Inserting the "Zheng Biao" (the Gnomon) in the hole at the center of the stone compass, and erecting the "You Biao" (sticks around the circumference), to observe the positions of the shadow projected by the Gnomon in the sun

西汉时期 漏壶 铭刻"千章铜漏一，重三十二斤""河平二年"(前 27)"四月造"，"千章"是县名
A clepsydra in the Western Han Dynasty with inscriptions of "Qian Zhang Bronze Water Clock, 32 Jin in Weight", "the 2nd year of He Ping" (27 B.C.), and "Produced in April". Qian Zhang was the name of a county

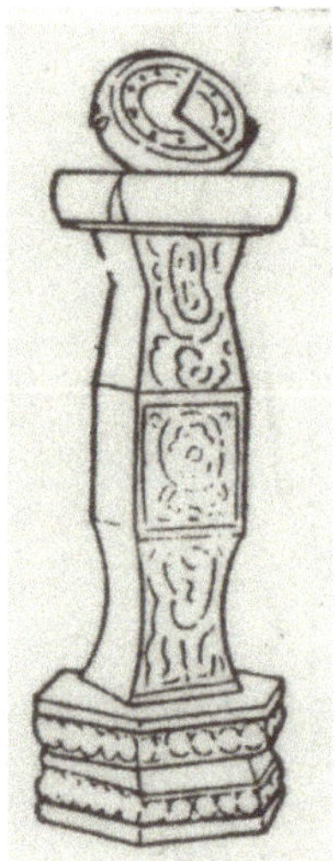

南宋时期 赤道式日晷图
A diagram of a sundial in the Southern Song Dynasty

西汉时期 日晷 用日晷测量白昼时刻，以调整漏壶的计时误差
A sundial in the Western Han Dynasty for gnomonic projection, to adjust the instrumental error in time reckoning by the water clock

1. 天文计时仪与漏刻

汉代的天文计时仪器具有很高的技术水平。如张衡（78 ～ 139）设计制造的“水运浑象仪”即以水为动力，是主要用于演示天体运动的仪器。它形象地说明宇宙的结构理论，兼有计时功能。浑象仪上标有二十四节气，并能反映太阳、月亮和五大行星在天体中的位置。根据太阳在浑象上的位置，不但能知道当时所处的节气，还可以大体知道所处的时刻。表明它已经具备了与机械钟表类似的功能。张衡制造的水运浑象仪，是现今见诸于记载的第一台用水力发动的天文仪器，对后代的影响很大。此后历代大多是在它的基础上发展出更复杂、更完善的天象表演仪器和天文钟的。此外，这期间的漏刻计时已达到一定的精度。据有关专家实验证明，汉代多采用二级补偿式浮箭漏供水壶，这样便可使漏壶水位保持相对稳定，计时精度可提高到日误差在 40 秒以内。南北朝时还发明了一种特殊类型的漏刻——秤漏。通过漏水的重量和体积来计算时间，即“漏水一升，秤重一斤，时经一刻”（一斤水对应的一“古刻”，相当于今天的 14.4 分钟）。计时的精度还可以随着秤的精度而提高。

唐宋两朝皆设有专司监、管漏刻和报时的官吏，还设有钟鼓楼按时敲钟击鼓以报时辰。唐时有一位多才多艺、聪明手巧的能人叫吕才，他制造了一种四柜式的漏刻。将漏壶的数量从晋代的三个增加到了四个。除吕才制四级补偿式浮箭漏之外，这时期秤漏仍十分流行。为了更进一步提高计时精度，宋代又研制成一种利用漫流法技术来保持水位恒定的计时器——莲花漏。宋代对计时器进行长期精心研究的，还有

东汉时期 天文学家张衡建立了宇宙论学说
Zhang Heng, an astrologer in the Eastern Han Dynasty, developed the theory of cosmology

水运浑象仪中计时部分的设想图
A diagram of time reckoning by applying the water-driven astronomical clock (Shui Yun Hun Xiang Yi)

东汉时期 张衡水运浑象仪模型
A model of Shui Yun Hun Xiang Yi, a water-driven astronomical instrument invented by Zhang Heng in the Eastern Han Dynasty

著名科学家沈括。他除制造了浑仪、浮漏、景表之外，还著有《浮漏仪》流传于世。该书成为中国古代论及漏刻最详细、最完整的一部重要文献，对后人了解漏刻结构及消除误差的各种措施提供了翔实可靠的文字资料。

唐宋时期民间常用的漏刻还有盂漏、田漏、船漏等。盂漏由大小两只铜盂组成，大盂盛水，小盂则放在大盂水面上，底部钻一针眼大小圆孔，利用小盂本身的重量作用让水涌入盂底孔内而逐渐下沉，小盂内水位逐渐提高直到小盂沉没。经实验证明，小盂经过 12 小时完全沉没，误差小于 15 分钟，已能满足民间计时的需要。

田漏是一种农家常用的计时器具。据记载，它是以日晷或圭表来校验泄水壶中一昼夜流出的水量是否与今日午至明日午圭表所示日影位置相合。说明田漏计时是以天文测时来校准的。

唐宋漏刻还被用在船舶上。诗人杜甫在《宿青草湖》中咏道：“洞庭犹在目，青草续为名。宿桨依农事，邮签报水程。”“邮签”即漏箭，行船时用以报时。从诗中还可以知道，当时已用漏刻来计算航程。当船工掌握了不同情况下的航速后，再根据航行的时间就可以计算出航程了。

张衡地动仪模型
A model of the seismograph invented by Zhang Heng

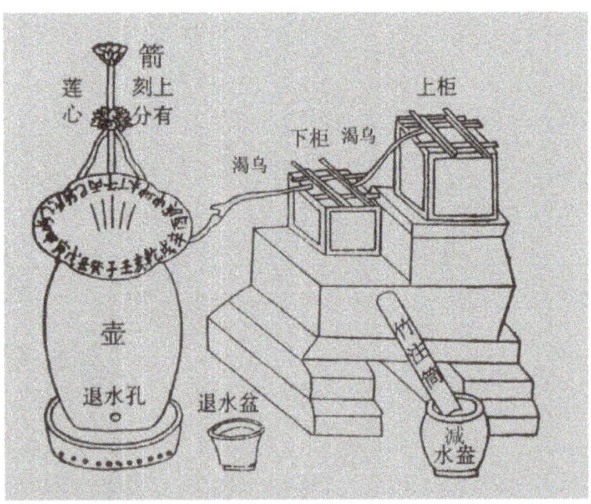

北宋时期 燕肃莲花漏示意图
A diagram of Yan Su Lian Hua Lou (Lotus Clepsydra) in the Northern Song Dynasty

唐代 吕才制造的四柜式漏刻图
A diagram of a four-partition Lou Ke (water clock) invented by Lu Cai in the Tang Dynasty

唐宋时期 民间使用的计时器——田漏
A Tian Lou (water clock) commonly used throughout the Tang and Song dynasties

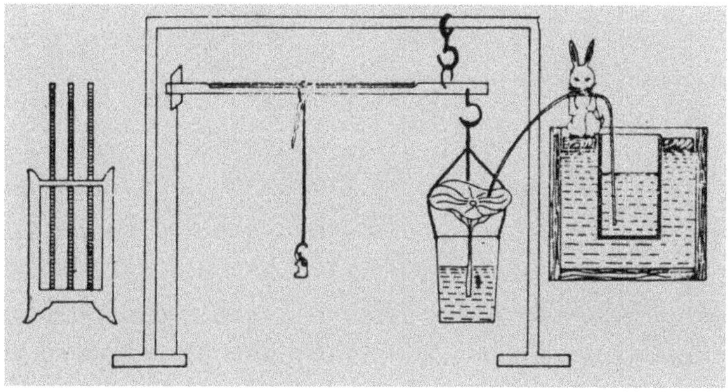

南北朝 道士李兰发明的秤漏 唐《初学记》记：李兰秤漏"漏水一升，秤重一斤，时经一刻"
A Cheng Lou (weighing clepsydras) invented by Li Lan, a Taoist priest in the Northern and Southern Dynasties. It was recorded in *Chu Xue Ji* in the Tang Dynasty that "on Li Lan's Cheng Lou, with 1 sheng of water dripping through, the scale showed a weight of 1 jin, indicating that 1 ke has been elapsed"

1. Devices for Celestial Observation and Time Reckoning

Devices for time keeping reached a high level in the Han Dynasty. Take Zhang Heng's (张衡) Shui Yun Hun Xiang Yi (水运浑象仪), for example. Designed by Zhang Heng (张衡, A.D.78-139), it was an armillary globe moved round using hydro energy, to demonstrate the movements of heavenly bodies. Such device illustrated the theories of cosmography, the structures of the universe. It was used for time keeping as well because it contained mechanisms like those of a clock. On this device the twenty-four solar terms like Jie Qi (节气, Seasonal Divisions) were marked, and relative positions of the sun, the moon, and of the five planets were shown when the device was moved round. Seeing from the perspective of the sun on the armillary globe, one would be able to read the solar terms and the time. Shui Yun Hun Xiang Yi (水运浑象仪) was the first one that was driven by hydro energy in the Chinese history, and had a great influence on devices of this type in generations to come. For example, in the Tang (唐朝) and Han (汉朝) dynasties some more complex and perfect armillary globes, as well as astronomical clocks were made, following Zhang Heng's designs.

Clepsydras known as Lou Ke (漏刻) during this period of time developed and provided a good method of calculating time. According to the findings of recent researches, in the Han Dynasty the two-grade compensating clepsydras with graduated floats known as Fu Jian Lou (浮箭漏) indicating the time elapsed became popular, and it was so precise that the average error within a day was less than 40 seconds. A special type of clepsydra called Cheng Lou (秤漏, weighing clepsydras) was invented in the Northern and Southern Dynasties (南北朝). When using this device for time keeping, the time was calculated by weighing the water in a vessel. When 1 sheng (升) of water, which weighed 1 jin (斤), flew from the vessel through the orifice, a quarter (刻) elapsed. Here the time unit ke (刻) was a concept in the ancient times (古刻), which was about 14.4 minutes in modern time reckoning. The mechanisms of this Cheng Lou (称漏) indicated that the precision of time keeping was associated with weighing and that it could be improved with the progress of weighing technologies.

In the Tang and Song dynasties there were officials whose duty was managing water clocks and reporting the time. There were Bell Towers (钟楼) and Drum Towers (鼓楼) for the announcement of time. In the Tang Dynasty, a man with versatile talents and skills named Lv Cai (吕才) designed and produced a four-step compensating clepsydra (浮箭漏) with graduated floats indicating the time elapsed. Together with the previous 3 of such apparatus made in the Jin Dynasty (晋代), the total reached four by then.

At the same time, Cheng Lou (秤漏) was still in common use. In the Song Dynasty, Lian Hua Lou (莲花漏, Lotus Clepsydra) was invented, which was a kind of clepsydra using the overflowing technology to guarantee the water level in the clepsydra changing evenly. This greatly improved the accuracy of time-keeping. In the Song Dynasty, among those who devoted to the research of time-keeping, there was a famous scientist named Shen Kuo (沈括). Besides making armillary spheres, clepsydras and sundials, Shen Kuo devoted much of his time to writing and most of his works became well known. In his book *Fu Lou Yi* (《浮漏仪》), research findings were recorded in details and in a comprehensive manner, offering explanations about clepsydras. Shen Kuo's works provided full and accurate descriptions for people of later ages to learn and understand the structures of clepsydras and technologies for eliminating errors.

Besides the clepsydras mentioned above, some other kinds of clepsydras such as Yu Lou (盂漏), Tian Lou (田漏) and Chuan Lou (船漏) were common in daily life throughout the Tang and Song times. Yu Lou (盂漏) composed two bronze Yu (盂, jar) of different sizes. The bigger jar was full of water and the smaller one floating on it. There was a tiny hole on the bottom of the smaller jar, the water was slowly flowing through it. After 12 hours the smaller jar would sink. Through experiments it was found that the error was less than 15 minutes. This accuracy was enough for the time reckoning in daily life.

Tian Lou (田漏) was usually used in peasants' families. Generally it was used together with a sundial (日晷) to cross-check and reach the exact time by comparing the time read from the clepsydra and that from the sundial. This showed that time keeping using Tian Lou depended on the astronomical clocks.

Chuan Lou (船漏) was a kind of clepsydra used on ships (chuan 船) for time announcing. If the shipman was able to tell the average speed of the ship, then the time elapsed shown on the clepsydra would enable him to calculate the voyage. Such a technology was brought into one of the poems, Staying by *Green Grass Lake* (《宿青草湖》) written by Du Fu (杜甫), a well-known poet. It read—Arriving at and staying by the Qingcao lakeside, while the Dongting Lake was still within sight.As you qian [邮签, in the poem referred to Chuan Lou (船漏)] showed me the voyage, I rest with my paddles in a peasant's house as a guest.

古代希腊的水钟，注满水后约6分钟漏完
A water clock in ancient Greece with the water dripping through in about 6 minutes

2. 计时机械的发展

继汉代张衡之后，在对天象的观测和仪器制造方面有显著进步的是唐宋时期。唐代天文学家和大地测量学家高僧一行，为编撰《大衍历》而组织了大规模全国性的天文大地测量。所用的"量天尺"即唐小尺，尺长24.5厘米。根据这次测量可换算出北极高变化1度，南北之间距离为129.22千米。唐开元十三年(725)，一行又和梁令瓒等人合作制成了水运浑天仪。浑天仪为铜质球形，球体浑象内列满了星宿，注水冲轮使它旋转，自转一周为一昼夜。外围安置两个圆环，上缀日月，每日西旋一周，日东行一度。365日沿黄道(太阳在天球中的视运动轨道)移动一周为一年；月标每昼夜回转一周，29日半沿白道(月球在天球中的视运动轨道)移动一周为一月。根据浑象上太阳的周日运动便可以得知当时的具体时刻。这台浑天仪最重要的改进是采用了自动报时装置。即"立木人二于地平上，其一前置鼓以候刻，至一刻自击之；其一前置钟以候辰(约合今2小时)，至一辰自撞之。"它的音响作用与近代机械钟表史上的自鸣钟作用相同。这台水运浑天仪既能表示天体运动，又能指示时间，是天文钟的前身。

宋代在水运浑天仪的制造方面又出现了一个新的高峰。元祐七年(1092)，由苏颂、韩公廉设计制造了一座规模宏大的水运仪象台，是一座兼具测天体运行、演示天象变化和报告时辰等多种功能的巨型天文机械装置。整座仪象台用同一套漏壶中泄出的水构成水力驱动系统。它包括提水、注水、回水机构和枢轮等部分，让水周而复始均匀流动，并驱动枢轮(原动轮)

运转。全部齿轮从6个传动到600个。枢轮直径11尺(约3.4米),顶部附设一组杠杆装置,相当于机械钟表中的擒纵轮机构,可控制枢轮作定速转动。报时装置称昼夜机轮,由传动齿轮与所有报时齿轮和浑象关联,是5层木阁式建筑。木阁中有158个木人,分别定时击打钟、鼓、铃、钲等不同的乐器和抱着时辰木牌循环出入自动报时。它是世界上最早制成的融测时、守时和报时为一体的综合性授时天文台。

北宋时期 天文学家苏颂
Su Song, an astrologer in the Northern Song Dynasty

2. The Improvement of Time Reckoning Devices

In the Tang and Song Dynasties, Chinese people made remarkable progress in the observation of heavenly bodies and in the production of planetarium. The famous astronomer and geodesist named Yi Xing (一行), who was also an eminent monk, organized a nationwide activity of measuring the earth which resulted in his compiling *Da Yan Li* (《大衍历》). The ruler used in this measurement was Liang Tian Chi (量天尺) which measured 24.5 cm. Based on the result of the measurement, it was calculated that the ratio between the latitudinal degree and the angular distance was 1°：129.22 km. In the 13th year of Kai Yuan (开元, A.D.725), Yi Xing cooperated with Liang Lingzan (梁令瓒) and some other collogues on producing a hydraulic armillary sphere known as Shui Yun Hun Tian Yin (水运浑天仪). It was made of bronze and composed of 3 layers. The inside layer, rotated by means of "flowing water", was full of marks of celestial bodies. It took a day to make one rotation. The middle layer, including latitudinal circles, the equatorial circle, and longitudinal circles, was fixed. The outside layer included two rotary circles: the ecliptic circle and the circle of the moon's orbit, on each there was a fixed mark representing the sun and the moon respectively. The ecliptic circle rotated clockwise. Everyday it moved one degree, and it took one year to make a rotation. It took 29.5 days to rotate around the moon's orbit, which was one lunar month of that

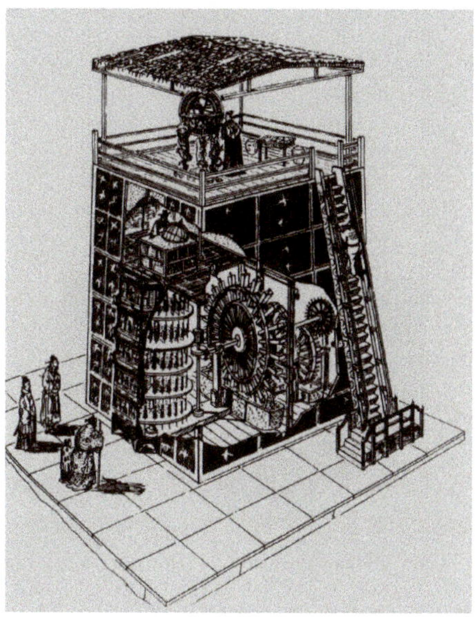

苏颂、韩公廉造水运仪象台是以水为动力，融测时、守时、报时为一体的天文钟
The Water-driven Astronomical Clock Tower invented by Su Song and Han Gonglian that integrated time reckoning, time keeping and time signaling

time. This armillary sphere demonstrated the motion of the sun, the moon, and the planets, and presented the correct time by showing the changing positions of the mark of the sun. With an automatic time announcing device installed, there appeared some improvements on the armillary sphere designed. There were two jacks on the armillary sphere, one of which would beat the drum every quarter and the other of which would beat the bell every Chen (辰, around 2 hours). Thus this device of armillary sphere was also known as a kind of striking clock (自鸣钟). Since the armillary sphere was associated with the time reckoning which was in turn associated with the motion of the celestial bodies, it would be claimed as the predecessor of modern astronomical clocks (天文钟).

In the Song Dynasty there appeared a new tide of constructing hydraulic armillary spheres. In the 7th year of Yuan You (元祐, A.D. 1092), Su Song (苏颂) and Han Gonglian (韩公廉) constructed a giant astronomical mechanical device which demonstrated the motion of celestial bodies and announced the time. This device was surmounted by a huge hydraulic armillary sphere, including a series of clepsydras, a water-circulating mechanism and a pivotal wheel. The clepsydras and the water-circulating mechanism allowed the pivotal wheel to rotate continuously a standard weight of circulating water running from the clepsydras. The pivotal wheel, 11 chi (尺) which was around 3.4 m in diameter, with a trip level movement like the escape wheel in modern clocks in function on top of it, rotated in uniformed motion, and made around 600 gears rotate. The time announcing device was in fact a five-floor tower, in which there were 158 jacks, some of which would play a certain kind of musical instrument and some others would appear with time-plate when the time was due. The mechanism in the time announcing device was working in correspondence with the rotating armillary sphere through the movement of transmission gears. This astronomical mechanical device was the earliest multi-functional time service device in the world that performed the function of time reckoning, time keeping, and time announcing.

浑象天球仪模型 系水运仪象台的一部分，仪下为地柜，上为经纬环规

A model of astronomical instrument that was a part of the Shui Yun Yi Xiang (water-driven astronomical clock) Tower, with the base and surrounding longitude-latitude hoops

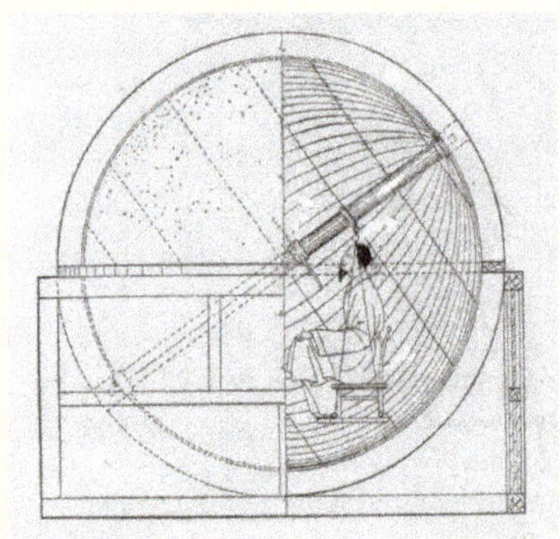

假天仪模型 下为方柜，柜上有经纬两种环规，布置星位坐标。人坐在球内便能形象地观看星空

A model of astronomical instrument, with the base and surrounding longitude-latitude hoops to display heavenly bodies on the coordinates. Sitting inside the instrument, one is able to visualize the star-dotted sky

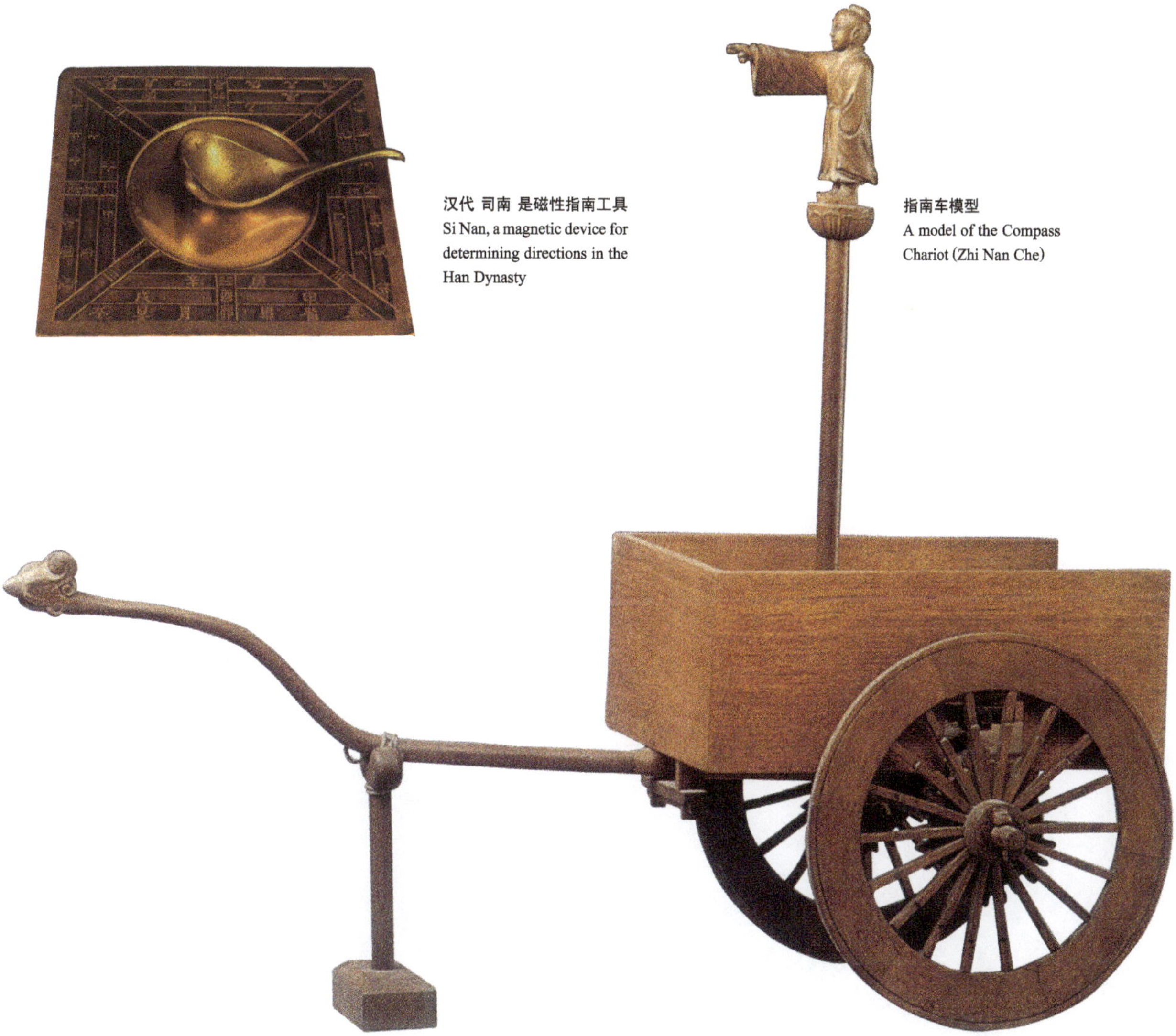

汉代 司南 是磁性指南工具
Si Nan, a magnetic device for determining directions in the Han Dynasty

指南车模型
A model of the Compass Chariot (Zhi Nan Che)

3. 指南车与记里鼓车

指南车和记里鼓车都是在皇帝出行时作为象征皇权的仪仗车辆。历代史书对这两种车的记叙很多，大部分只对其功能和外形作描绘。直至宋代才对车的内部结构、齿轮和传动系统有了较详细的叙述。现代科技史家王振铎先生对车做了深入研究，并复制有模型，今陈列在国家博物馆。

(1)指南车

据记载，东汉时张衡就曾制造过指南车。其后三国、晋、南北朝先后都有人再作尝试，并不断加以改进。唐代的马待封、金公立也制成过指南车。对指南车和记里鼓车叙述最详的是《宋史·舆服志》。书中分别对燕肃(1027)和吴德(1107)所造指南车的规范、结构、齿轮数量和它们的直径、周长、齿距等都做了详细的叙述。

指南车的结构并不复杂，但机械传动和自动控制原理却十分巧妙。它是由与指示方向的木人相连接的中心大齿轮(主齿轮)及与车子左右车辕相连接的左右小齿轮等组合而成的。关键在于控制左右小齿轮在车行驶状态中的离合。木人指南原理类似今天汽车方向盘的逆向传动原理，汽车直行，方向盘不动；转动方向盘经过齿轮系统使车轮转弯，改变行车方向。而指南车则是通过车轮转弯，经过齿轮系传动，拨动木人改变方向。指南车的发明是前人在机械设计、制造方面智慧的结晶。

3. The Compass Chariot (指南车, Zhi Nan Che) and the Drum Chariot (记里鼓车, Ji Li Gu Che)

The Compass Chariot (literally South-Pointing Vehicle) and Drum Chariot (literally Mileage-Reckoning Vehicle) in ancient China were symbols of imperial power that followed the emperor on a sightseeing trip. Records about those "vehicles" are numerous in historical works. But most of these records were about the general function and contours. It was not until the Song Dynasty (宋代) that records about the inner structures, gears, and transmission systems were graphically described. In the modern times, an expert in history of science, Wang Zhenduo (王振铎), made an in-depth research on the vehicles and reconstructed a model, which is now an exhibit in the National Museum of China.

(1) The Compass Chariot

According to historical records, as early as in the Eastern Han Dynasty (东汉), Zhang Heng (张衡) once built a south-pointing chariot, namely, Zhi Nan Che (指南车) to indicate directions. Later in the times of the Three Kingdoms (三国), Jin (晋朝), and in the Northern and Southern Dynasties (南北朝), more inventors exerted efforts on trials and producing such vehicles with improved technologies. In the Tang Dynasty, Ma Daifeng (马待封) and Jin Gongli (金公立) also successively produced such a chariot. The most detailed records about Zhi Nan Che can be found in *Song Shi / Yu Fu Zhi* (《宋史 · 舆服志》), in which the regularity, inner structures, number of gears with their diameters and perimeters were described specifically on Zhi Nan Che built by Yan Su (燕肃, A.D.1027) and Wu De (吴德, A.D.1107) respectively.

The structures of Zhi Nan Che were actually not complicated. Nevertheless, the mechanisms for transmission and principles for automatic control were very ingenious. The main parts of the vehicle were the transmission gears that joined the gears rotating the jack-worker and the gears rotated by the right and left axletrees, and the key point to make the vehicle work properly was to control the engagement of the transmission gears to the gears rotated by the axletrees. Such a principle of the workings of the vehicle was something like that of the steering wheel, gear box and the clutch in a modern car. The moving of the steering wheel is in correspondence with the turning of directions. As for Zhi Nan Che, when the vehicle made a turn of a certain degree in its direction, through the work of the gears, the gear rotating the jack-worker would move the same degree accordingly. This invention in machinery designing and construction was the cream of crops of ancient Chinese people's wisdom.

指南车在战争中起到重要的作用

The significant role the Compass Chariot played in battles

(2)记里鼓车

记里鼓车又叫“记道车”。汉代刘歆著《西京杂记》中有“记道车，驾四，中道”的记载。汉画像石有“鼓车”图。唐代以前车仅一层，唐以后分为两层：即行一里下层木人击鼓，行十里上层木人击镯。据记载，汉代以后历代制造记里鼓车者也不乏其人，但文字皆过于简略。至宋代，车的结构、原理及制造方法均已失传。北宋天圣五年(1027)卢道隆重新设计制造成车。《宋史·舆服志》则详细记述了车的结构、尺寸、规范等。记里鼓车是我国古代对齿轮系运用的典范。它的记程功能是由齿轮系完成的：将两个齿数相同的齿轮，中间嵌入一个中轮，车便能按同一速度和同一方向运转，整个齿轮系与车轮同行同止。车上还有一套减速齿轮系，始终与车轮同时转动，其最末一只齿轮轴在车行一里时中平轮正好回转一周，经机械系统作用，车上的拨子就拨动下一层木人击鼓一次；车行10里时，上平轮转一周，其上的拨子就拨动上一层木人击镯一次。今天汽车的里程表每行驶1千米便转动1个数码，其原理与古代记里鼓车相似。

汉代 画像砖上的“鼓车”图
A diagram of the Drum Chariot drawn on bricks in the Han Dynasty

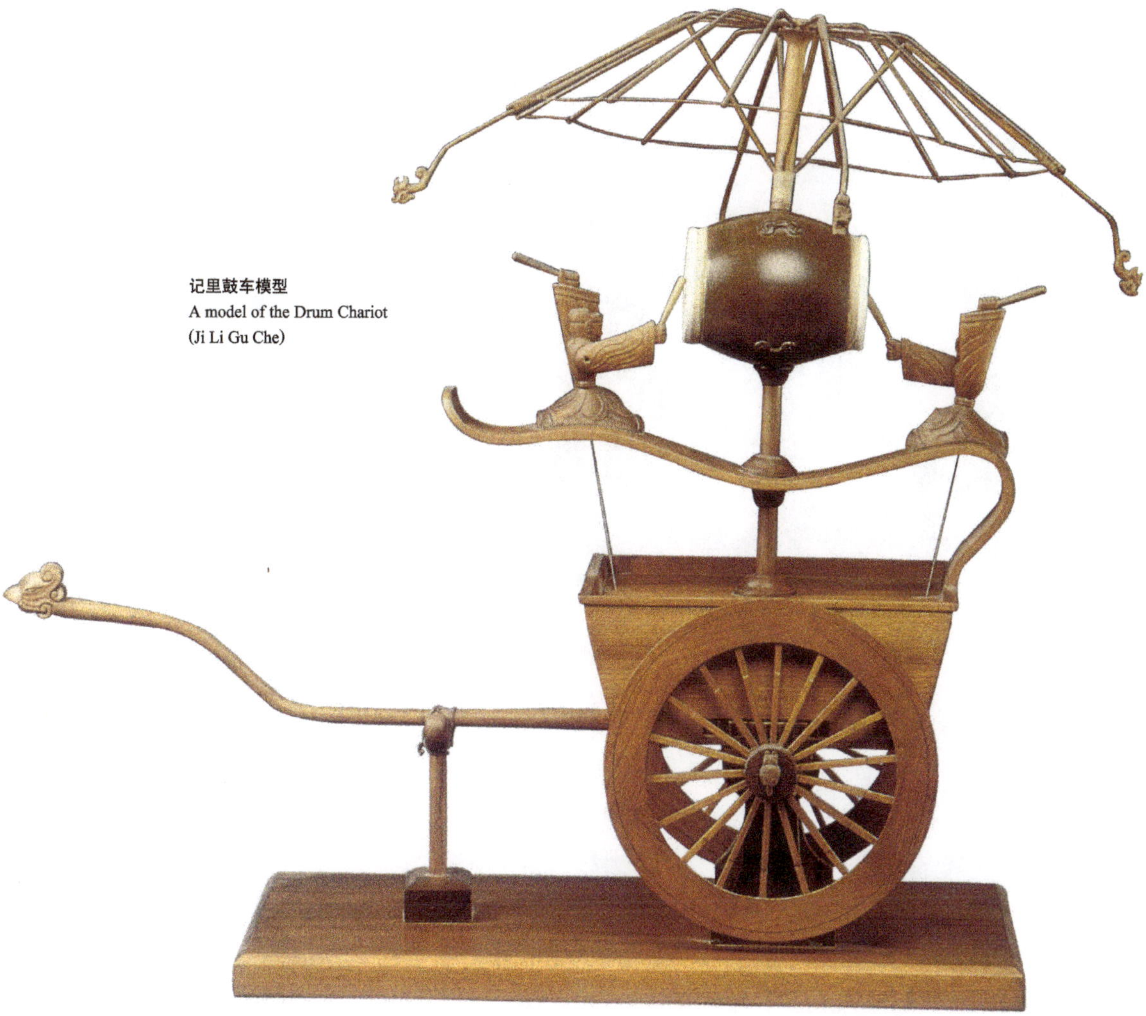

记里鼓车模型
A model of the Drum Chariot (Ji Li Gu Che)

(2) The Drum Chariot

The Drum Chariot, in Chinese, Ji Li Gu Che (记里鼓车) or Ji Dao Che (记道车), was recorded in *Xi Jing Za Ji* (《西京杂记》) written by Liu Xin (刘歆) in the Han Dynasty. On some stone inscriptions of the Han Dynasty we can see the patterns of such vehicles. Since the Tang Dynasty, the mileage reckoning drum vehicle had developed into a two-storied vehicle. Jack-workers on the first story would beat drums per li (里, 500 m), and those on the second story would beat zhuo (镯, a kind of ancient Chinese musical instrument) every 10 li (里). A reading of historical works gives us a belief that after the Han Dynasty people who were able to build such mileage reckoning drum vehicles were not few, but records in these historical works were all very sketchy. By the times of the Song Dynasty, the structures and pertinent assembly methods had been lost. But in the 5th of Tian Sheng (天圣, A.D.1027) of the Northern Song, Lu Daolong (卢道隆) re-designed and re-built on his own this type of drum vehicle. The structure, size and regularity of Lu Daolong's vehicle were recorded in *Song Shi / Yu Fu Zhi* (《宋史・舆服志》). The Drum Chariot was a symbol of the skillful application of gears in ancient China. The function of mileage reckoning was accomplished through the arrangement of gears: the wheels drove a device through a series of transmission gears to make sure that as the wheels were turning a certain revolutions, which meant that the vehicle had moved a certain distance in terms of 1 or 10 li (里), and that the transmission gears activated the device and thus made the jack-workers beat the drums or zhuo. There was also a breaking mechanism attached to the wheels that synchronized the turning of the wheels. The principles of mileage reckoning drum vehicles were similar to those of odometers in modern cars.

指南车和记里鼓车都作皇帝出行时的仪仗车队
Both the Compass Chariot and the Drum Chariot were among the royal fleet of vehicles as part of the Guards of Honor

4.大气湿度的测量

“湿易雨”，大气的湿度和晴雨的关系极为密切。为了掌握气象变化的规律，古人十分注意观燥湿的变化，并不断地发明和改进测验大气燥湿以预知晴雨的方法。汉代文献记载：“悬羽与炭，而知燥湿之气。”古人认为，阳气为火，阴气为水。“水胜，故夏至湿。火胜，故冬至燥。燥故炭轻，湿故炭重。”早期的湿度计，都是利用天平的失衡来观测空气的燥湿，即把两个重量相等而吸湿性不同的物体，如炭和土，分别挂在天平的两端。木炭吸水性大于土，当空气干燥时，木炭的水分很容易蒸发而变轻。当空气湿度增大时，木炭吸收了空气中的水分而变重。根据天平的倾斜变化，测定空气湿度并推测阴晴。

空气的燥湿度还会影响琴弦的张力变化，从而使弹奏出来的音调缓急发生变化。故有“风雨之变，可以音律知也”和“天且雨……琴弦缓”等之说。古代做琴弦常常用动物的筋、革、肠等，这些材料质地都属于动物蛋白纤维，有一定的吸湿性。风雨来临之前，空气湿度增大，琴弦会因吸湿而增长，奏出来的音调也就变低，所以又有“音律知天”之说。这中间已经孕育着悬弦式湿度计的基本原理，成为后代制造毛发湿度计的依据。

汉代 悬土炭测量大气湿度的变化
In the Han Dynasty, there existed a technique of suspending pieces of charcoal to measure humidity and its variations

4. The Measurement of Humidity

There was an old saying in ancient China, "high humidity often leads to a rain" (湿易雨). That humidity was directly related to the rain had been a common sense. In order to master the regularity of the changes of weather, ancient Chinese people paid much attention to the changes of humidity, and continuously created methods to measure it. Ancient Chinese people thought that "fire (dryness) stems from yang qi (阳气), and water (humidity) stems from yin qi (阴气)."The early humidity device was made according to principles of balances. Records in the Han Dynasty (汉代) read that "an equal-arm balance with a lump of charcoal suspended at one end of the beam and a piece of feather suspended at the other end could be used as the simplest humidity detector. " As a kind of humidity absorbing material, the lump of charcoal would be heavier when the humidity increased, and it would be lighter when the air was getting drier. Therefore, by observing the angle of the inclination of the beam, the humidity could be detected. Ancient Chinese people were also forecasting the weather by referring to the change of the humidity. The change of humidity would affect the pulling force of strings, thus changing the pitch of a stringed musical instrument. So in ancient China it was said, "the strings' vibration frequency is lower when it is going to rain", and "the change of weather can be forecasted through the change of Yin Lv (音律, the Chinese term for pitch)". In ancient China strings were made of intestines from animals, which were protein fibers in nature, and were humidity-absorbing materials. Before the rain, the degree of humidity would increase, and the strings would be "stretched" longer than usual. At this time the pitch of the musical instrument would sound lower. Ancient Chinese people's understanding of the relationship between the change of the weather and the change of the pitch inspired later generations to invent hair humidity detector.

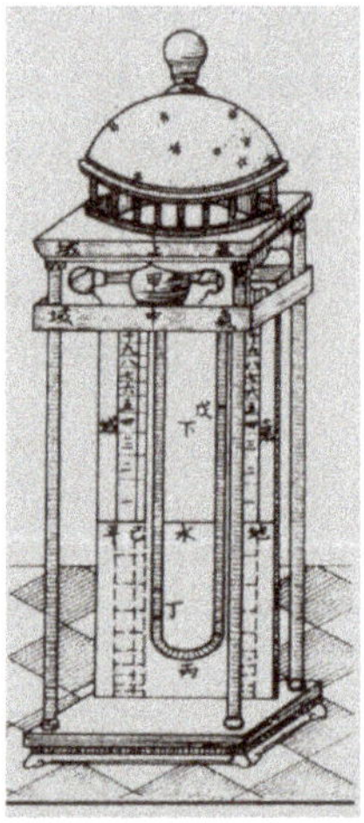

清初 比利时传教士南怀仁(1659 年来到中国)制作的毛发湿度计

During the early years of the Qing Dynasty, Ferdinand Verbiest, a missionary from Belgium who came to China in 1659, invented a hair hygrograph

据记载,中国最迟从 11 世纪起就记录冬至后九个九天当中每日的天气,故有"数九寒天"之说

According to historic records, ever since the 11th century, there have been daily accounts of the weather during the nine nine-day periods after Winter Solstice, hence there is a saying of "counting the nine cold days nine times"

汉代 观察琴弦的变化，推知大气的燥湿
In the Han Dynasty, people deduced the humidity from observing the changes in the strings of musical instruments

汉代 画像砖上的相风鸟
A device for detecting wind directions drawn on bricks in the Han Dynasty

5.风向、风速的测量

刮风下雨是十分常见的自然现象。人们在长期的生产实践中，对风向和风速的认识从感性到理性，讴歌它、观察它、测量它。甲骨文中已有祭祀东、南、西、北四方风神的记录，并对风向设定专门的名称。

早在先秦时期已制造了风向仪——伣。汉代文献《淮南子》说："伣之见风也，无须臾之间定矣。""伣又称谍，取侦察风向之意。伣的结构简单：立一风杆，杆上系丝帛做成的长条形"旗"，称"示风器"，器上系一小铃，风吹铃响。设有专门的观测者监听铃声并看旗被风吹动的方向以报风向。西汉时风向器除"伣"之外，还有相风鸟。"铸铜凤，高五尺，饰黄金，栖屋上，下有转枢，向风若翔。"铜凤凰下部有转枢，插在一个圆槽内，可随风转动，使凤凰头部总是指着风吹来的方向。"下有转枢"暗示转枢必与下一层的转动机件相连，可能装有一种记风速的器件，是风速计的先驱。东汉至三国改用木鸟作风向仪，更为轻便，并且使用也更普遍了。

唐代科学家李淳风(602～670)在他的著作《乙巳占・占风远近》中根据风对树产生的力来估测风速。"树叶微动，风速约十里；树叶沙沙作响，风速则日行百里；树枝摇，二百里；堕叶，三百里；折小枝，四百里……"再根据树的摇动情况定出风级："一级叶动，二级鸣条，三级摇枝，四级坠叶，五级折枝，六级折大枝，七级飞沙石，八级拔大树及根。"外加"无风""和风"共十级。与近代各国对风等级的划分相差不多。

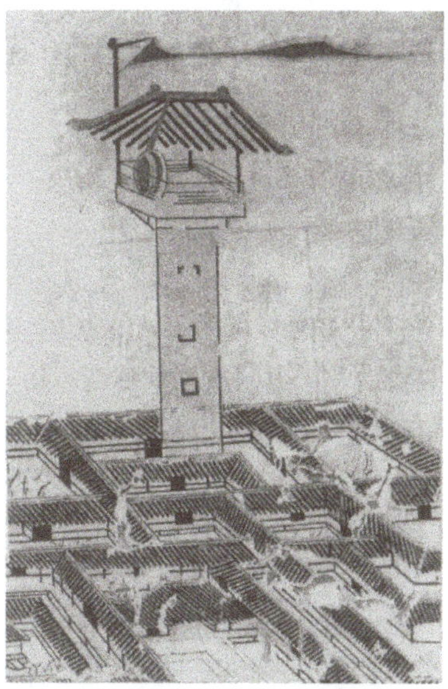

汉代 画像砖上的"伣"
A device for detecting wind directions drawn on bricks in the Han Dynasty

5. The Observation of Wind Directions and the Determination of Wind Speed

In China, the observation and research on wind started long time ago. As early as it was the Shang Dynasty (商朝) Chinese people offered sacrifices to wind-gods (风神) in the four directions, north, south, east and west, and gave names to wind directions, which can be seen from Jia Gu Wen (甲骨文), the earliest Chinese characters and inscriptions carved on bones or tortoise shells.

In the Qin Dynasty (秦朝), Chinese people started to make a kind of wind vanes known as Qian (俔). In *Huai Nan Zi* (《淮南子》) which is a famous historic collections compiled in the Han Dynasty, it was recorded that "Qian was used for the observation of wind and the wind direction could be told in an instant". Qian was also called Die (谍), meaning it was used for the detection of the direction of wind. Qian was quite simple a structure: on the top of a long pole a small bell and a flag-shaped silk ribbon were tied. When the wind blew, the bell would ring, and the wind observer would record the wind direction and report it right away. In the time of the Western Han (西汉), Xiang Feng Niao (相风鸟), a phoenix-like bronze object (铜凤) with a height of 5 chi (尺) was invented for detecting wind directions. It was cast in bronze and decorated with gold. "When installed on the roof of a house, it would turn on its axis and direct its head towards the direction the wind comes". In the Eastern Han Dynasty (东汉), Xiang Feng Niao was made of wood, and thus became more popular.

Li Chunfeng (李淳风, A.D.602-670), a famous scientist in the Tang Dynasty (唐代), put forward a way of reckoning wind speed in his works entitled *Yi Si Zhan / Zhan Feng Yuan Jin* (《乙巳占・占风远近》). In his book he suggested ways of determining the scale of wind speed, "The force 1 wind will sway leaves of trees; the force 2 wind will ring twigs; the force 3 wind will rock branches; the force 4 wind will blow leaves down; the force 5 wind will break branches; the force 6 wind will break strong branches; the force 7 wind will make sands fly up; and the force 8 wind will up-root trees. In addition there were two scales labeled as "calm" and "breeze". According to his array of scales, we can take him as the first one who offered the scales of wind speed in the world.

装置在宫殿屋顶上的相风鸟图
A diagram of a phoenix-like object for detecting wind directions mounted on the roof of the palace

北宋 相风鸟
A diagram of a phoenix-like object for detecting wind directions in the Northern Song Dynasty

一级叶动
Force 1—the wind that sways leaves of trees

三级摇枝
Force 3—the wind that rocks branches

五级折枝
Force 5—the wind that breaks branches

六级折大枝
Force 6—the wind that breaks big branches

唐代 科学家李淳风把风速划分为 10 级
Li Chunfeng, a famous scientist in the Tang Dynasty, categorized wind speed and force into 10 scales

THE HISTORY  OF ANCIENT CHINES

EASURES AND WEIGHTS

10

THE HISTORY OF ANCIENT CHINESE MEASURES AND WEIGHTS

中国古代计量史

十、元明清时期的度量衡

Chapter Ten

Measures and Weights in the Yuan, Ming and Qing Dynasties

1. 元代的度量衡

元代（1206～1368）国土幅员广阔，交通发达。海外贸易也得到发展，与日本、朝鲜、印度以及波斯湾沿岸各地都有交往，波斯人和阿拉伯人到中国来的就相当多。中国的三大发明火药、罗盘、印刷术都在此时经阿拉伯传入西欧。为了便于与各方面、各地区商贸往来，政府采取了各种措施，其中包括在秤锤上铸（刻）不同文字：汉文、回鹘蒙文、八思巴文和波斯文。不同民族、不同国家的文字在小小的秤锤上同时出现，除加强了各民族之间语言文化的交流外，同时也是为了适应外国人到中国经商的需要。

元立国后，基本上沿用了宋代原有的典章制度。宋代度量衡由国家规定专管商务贸易的太府寺掌管，禁止民间私造。《元典章》中规定："凡斛斗秤尺，须行使印烙。官降法物。"还规定凡改年号即铸造器具颁至全国。这一点在元代秤锤上反映得十分突出。迄今收藏在各地博物馆的元代秤锤甚多，展览的就有 300 余件，其中有年号的共 247 件。

元代秤锤铭文的内容除包括铸造的年代、地名、长官等外，有些锤上还铸有锤重和最大秤量，反映出元代杆秤已发展到一个成熟的阶段。通过对秤锤和有自重铭文的银铤等各种实物的实测，可以推证元代 1 斤约为 608 克。

元代一尺多长？一升容积是多少？至今尚未得到确切可靠的历史资料。经多方考证，也只能给出一个参考数值，即 1 尺约 35 厘米，1 升约 1003 毫升。

元代　铜砣　一面铸"大德八年"（1304）"大都路造"，另一面铸"五十五斤秤"和八思巴文"二斤锤"。其余四面分别铸回鹘蒙文和波斯文"五十五斤秤""二斤锤"。重 1275 克

A bronze sliding weight, 1, 275 g in weight, in the Yuan Dynasty with inscriptions on one side as "The 8th Year of Da De" (1304) and "Da Du Lu Zao"; on the other side as "For a 55-jin Steelyard" and "A 2-Jin Sliding Weight"; on the other 4 sides, as "For a 55-Jin Steelyard" and "A 2-Jin Sliding Weight" in Mongolian and Persian respectively

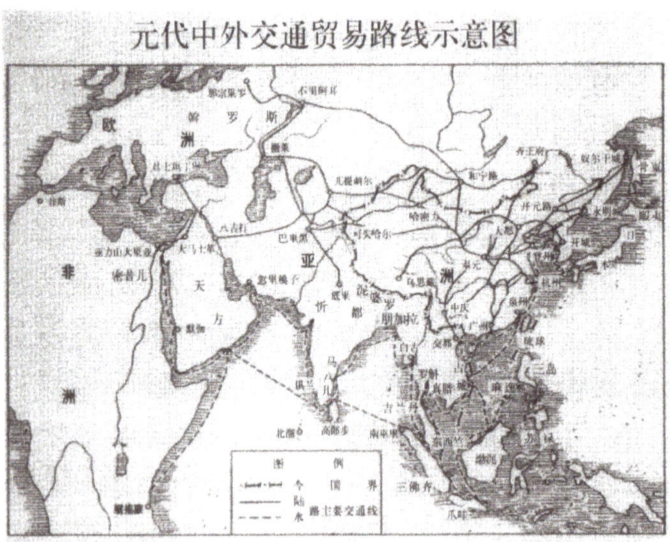

元代中外贸易路线示意图　元代海上贸易发达，曾与数以十计的国家和地区建立了海上贸易关系

An exhibition of foreign trade in the Yuan Dynasty, foreign trade with scores of countries by sea was flourishing in the Yuan Dynasty

1. The Measuring and Weighing System in the Yuan Dynasty

In the Yuan Dynasty (元代, A.D. 1206-1368), China was a country with a vast territory. The transportation and communication were progressive. The relationships with countries such as Japan, Korea, India, Persia, etc. were in very good terms. Overseas trade developed well, and in many Chinese cities in this period there were Persian and Arabian merchants doing businesses there. During this period of time, three of China's greatest inventions, namely, printing, gunpowder and compass started to spread westwards to Europe via the Arabian area. In order to facilitate trades between different areas and with foreign countries, the central government took various measures, such as engraving words in different languages on weights. This not only met the demands from foreigners doing trades in China, but also enhanced the exchange of language and culture.

Following the systems set in the Song Dynasty (宋代) as far as institutions were concerned, the government also strictly controlled measuring and weighing apparatus in the Yuan Dynasty. It was stipulated in *Yuan Dian Zhang* (《元典章》) that "such measuring and weighing apparatus as hu (斛), dou (斗), cheng (秤) and chi (尺) must be stamped to represent legality. They are legalized subjects that must only be distributed by the government". Furthermore, stipulated in *Yuan Dian Zhang* was that whenever there was a new title of an emperor's reign, measuring and weighing apparatus with the new title must be cast and distributed around the country. In the museums all over China, there are more than 300 pieces of weights made in the Yuan Dynasty collected for exhibition, among which there are 247 pieces engraved with the titles. All these show that the implementation of steelyards in the Yuan Dynasty

元代 铜砣 六面用四种文字铸铭文，汉字为"大德七年"(1303)"大都路造三十五斤秤"。重 520 克

A bronze sliding weight, 520 g in weight, in the Yuan Dynasty with inscriptions on the 6 sides in 4 languages.The inscriptions in Chinese read "The 7th Year of Da De" (1303) and "Da Du Lu Zao for a 35-jin Steelyard"

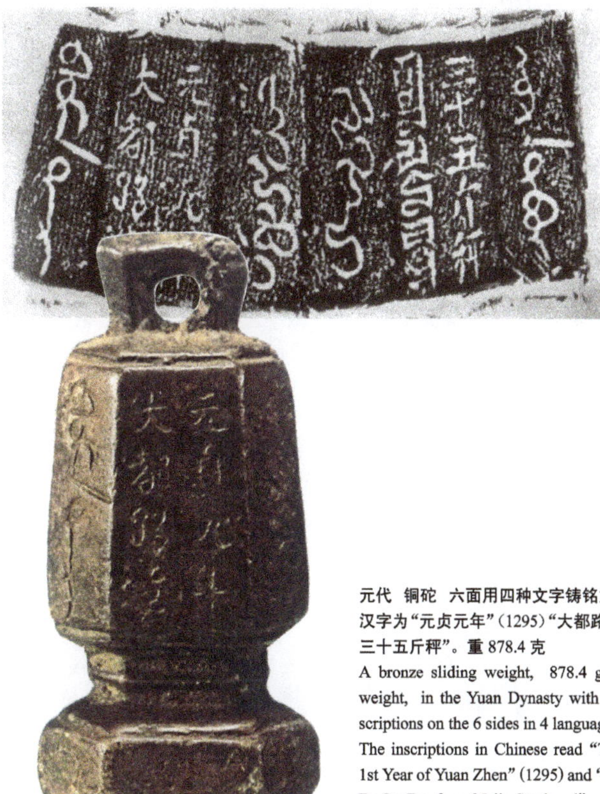

元代 铜砣 六面用四种文字铸铭文，汉字为"元贞元年"(1295)"大都路造三十五斤秤"。重 878.4 克

A bronze sliding weight, 878.4 g in weight, in the Yuan Dynasty with inscriptions on the 6 sides in 4 languages. The inscriptions in Chinese read "The 1st Year of Yuan Zhen" (1295) and "Da Du Lu Zao for a 35-jin Steelyard"

元代 铁砣 铸阳文"天历贰年"(1329)

An iron sliding weight in the Yuan Dynasty with characters cut in relief as "The 2nd Year of Tian Li" (1329)

became mature. Through weighing those weights and silver ingots engraved with maximum weights, we can make a judgment that 1 jin of the Yuan Dynasty was equal to about 608 g.

Due to the absence of reliable and unambiguous historical data, the precise length of Yuan Chi (尺, ruler) and the capacity of Yuan Sheng (升) remain unknown. However, our researches help to offer readers the following reference: the length of 1 Yuan Chi was about 35 cm, and the capacity of 1 Yuan Sheng was approximately 1,003 ml.

元代 铜砣 铸铭文"至元二年"(1336)
An iron sliding weight in the Yuan Dynasty cast with inscriptions as " The 2nd Year of Zhi Yuan (1336) "

元代 铜砣 铸铭文"大德七年(1303)造"
A bronze sliding weight in Yuan Dynasty cast with inscriptions of "Made in the 7th Year (1303) of Da De"

元代铸造了各种银锭，统称"元宝"，上面多标明自重，经过对定量秤砣和有自重铭文的元宝等实测后，综合分析得元代 1 斤约合 608 克
There were various types of silver ingots cast in the Yuan Dynasty known as "Yuan Bao" mostly with inscriptions of their weights. Conprehensive analyses, after cross-checking with the actual weights of the weights and the inscriptions on them, were reached that 1 jin in the Yuan Dynasty was about 608 g in weight

2. 明清时期的度量衡

公元 1368 年朱元璋推翻了元朝，建立了汉族统治的最后一个封建王朝——明朝(1368 ～ 1644)。随后是由满族建立的清朝(1616 ～ 1911)。明末清初，西方资本主义文明刚刚兴起时，中国还保持着“天朝上国”的神圣威严，仍然是矗立在东方的一个古老巨人。

明、清(公元 1840 年以前)是中国商业发展的重大转折时期。在国内商品经济比较发达的地区，稀疏地出现了资本主义萌芽。在集镇和城市，商业进一步发达，并且开展了对外贸易。但是在郑和“七下西洋”(1405 ～ 1433)之后，又实行了闭关锁国政策，使刚刚处于萌芽状态的资本主义商品经济受到了致命的打击，这一段时期阻碍了中国经济发展的进程。

2. The Measuring and Weighing System in the Ming and Qing Dynasties

In the year 1368, Zhu Yuanzhang (朱元璋) assumed the imperial title and founded the Ming Dynasty (明朝, A.D.1368-1644), which was the last one of the Han ethnic group in the Chinese history, followed by the Qing Dynasty (A.D.1616-1911) that was founded by the Manchu ethnic group (满族). During the late Ming and early Qing periods, when the capitalist civilization was on the upgrade in the western world, China, a strong empire with a long history, was still “an aged giant in the East” with a divine awesomeness.

Nevertheless, the development of commerce in China made an important turn during the transitional period from the Ming Dynasty to the Qing Dynasty (prior to the year 1840). In the areas where commerce developed well, nascent capitalism began to grow. In some cities foreign trade was becoming prosperous. But after the well-known “Seven Times of Voyaging to the Western Oceans (七下西洋, A.D.1405-1433) under the command of Zheng He (郑和), the regime adopted a close-door policy, which hindered the development of capitalism in its early stage in China, which in turn slowed down greatly the development of the Chinese economy for a period of time.

明代 嘉靖(1522 ～ 1566)牙尺 长 32 厘米
An ivory ruler, 32 cm in length, made in Jia Jing (1522-1566) of the Ming Dynasty

明代 牙尺 长 35.8 厘米
An ivory ruler, 35.8 cm in length, made in the Ming Dynasty

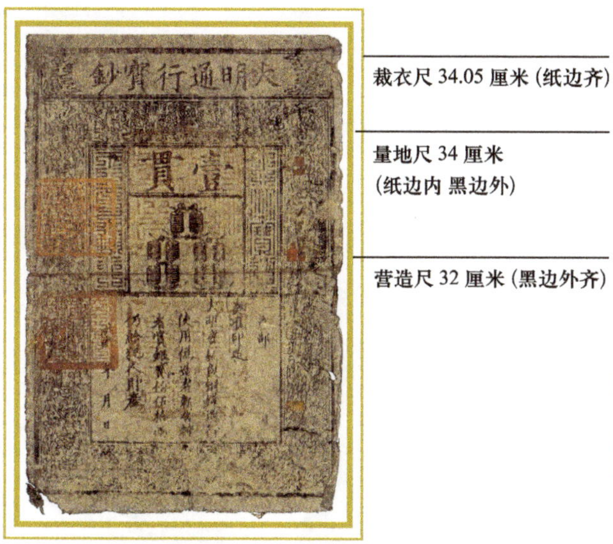

明代科学家朱载堉，为研究乐律而探讨古尺与明代尺度值之比。又以有明确尺寸规定的“大明宝钞”纸币比较明代各种日常用尺。今实测较完整的大明宝钞 39 张，求得营造尺长 32 厘米
In his research on temperament, Zhu Zaiyu, a scientist in the Ming Dynasty, explored the ratio between the lengths of ancient rulers and those made in the Ming Dynasty. He also compared the lengths of rulers commonly used, with the help of the banknotes “Da Ming Bao Chao” with regulations for the size. After measuring 39 “Da Ming Bao Chao” banknotes, the length of Ying Zao Chi was found as 32 cm

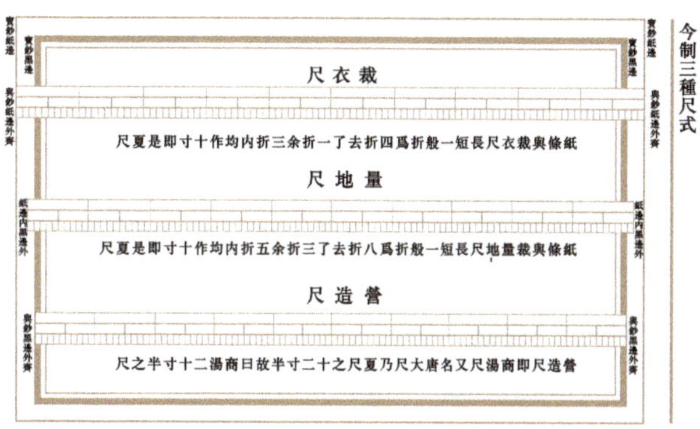

根据朱载堉绘三种尺式图复制
A diagram of replicated rulers in accordance with Zhu Zaiyu

(1)资本主义商品经济的发展与明代度量衡

明代前期,农民已更多地从事农业商品的生产,从而刺激了商品经济的发展,以致部分地主也开始对"工商末业"大感兴趣。明朝政府对手工业的控制也在逐渐削弱,一些地区的手工业部门中,突破了封建生产关系而出现了具有资本主义性质的雇佣关系。商业的繁荣促进了货币的更广泛使用。由于铜钱值小又不便携带,洪武八年(1375)政府发行纸币——大明宝钞。规定其"高一尺,广六寸"。经实测保存完好的大明宝钞,得知明代的营造尺长32厘米,与所见保存完好,制作精美的明代尺实测数值也相合。

大明宝钞发行后不久,因发行数额不断膨胀以致信用丧尽,民间交易又出现了用白银作货币。白银大量在市上作为流通货币,大交易用银锭,小买卖用碎银子,故称金银的戥秤广泛使用。此外木杆秤在民间也十分普遍。今见一件保存完好的瓷质秤锤,顶部有不等边六面体鼻纽,用以系绳,其中有两面书写"金玉""富贵",顶端台面上书写"万历肆拾柒年(1619)孟冬月吉呈制"。权身也呈六面体,每一面写一个字,连成"公平交易正直"。在明代之前,中国历代度量衡器上的铭文,多为反映与统治阶级政权相关的内容,威严肃穆,而瓷权上的铭文却一改过去的风格,出现了代表资本主义经营思想的内容。还有的砝码上刻有商号之间相互校验的铭文。随着资本主义萌芽的产生,过去一直掌握在统治阶级手中,主要用来征收赋税的度量衡器具,逐渐受到商人们的青睐,成为保证公平交易的工具。

从所见砝码和相关器物中,推算出明代1斤约596.8克,1升的容积约1035毫升,均是前代的延续。

明代 铜砝码 刻铭文"叁两"。重108克
A bronze weight, 108 g in weight, in the Ming Dynasty cast with inscriptions of "3 Liang"

明代 铁砣 铸阳文"洪武"(1368～1398)"官造"
An iron sliding weight in the Ming Dynasty with characters cut in relief as "Hong Wu" (1368-1398) and "Guan Zao"(Officially Produced and Promulgated)

明代 铜砣 铸莲花云纹
A bronze sliding weight in the Ming Dynasty cast with lotus designs

明代 盒装式铜砝码 刻铭文"崇祯丁丑年置"(1637)"贰拾两"，总重 689.7 克
Bronze weights made in the Ming Dynasty with carred inscriptions as "Made in the Year of Ding Chou in Chong Zhen"(1637) and " 20 Liang" They are 689.7g in total

(1) The Relationship Between the Capitalist Commodity and the Measuring and Weighing System in the Ming Dynasty

In the early Ming times, more and more peasants started to engage in the commodity production, thus helping the development of the commodity economy. In some areas, even landlords began to be interested in "the last choice as trades and handicraft industry and commerce". As the governmental control over the handicraft industry was gradually loosening, in some areas capitalism started to emerge. The prosperity in commerce was a strong push factor on the distribution of currency. Since copper coins, with small values, were becoming more and more inconvenient in the growing trades, the central government in the 8th year of Hong Wu (洪武, 1375) issued paper money named the Da Ming Bao Chao (大明宝钞) with the size of "1 chi (尺) in length and 6 cun (寸) in width". Through measuring those intact notes, we found that 1 chi of the Ming times was 32cm, which was in coincidence with the value calculated through measuring intact rulers made in the Ming Dynasty.

Because the issuance of large quantity of such notes caused losses in the credibility, silver began to replace them as a medium of exchange in social lives. The use of the precious metal led to the popularity of Deng Cheng (戥秤, steelyard), a weighing apparatus

明代 铜砝码 刻铭文"崇祯丁丑年置"(1637)"贰拾伍两"，是由官方颁发，商贾间相互校准通行的标准器。重 928.4 克
A bronze weight, 928.4 g in weight, in the Ming Dynasty cast with inscriptions of "The Year of Ding Chou in Chong Zhen" (1637) and "25 Liang", which was promulgated by the government as the standard for calibration among merchants

明代 铜砣 铸铭文"建文元年"(1399)"应天府(今南京)造"
A bronze sliding weight in Ming Dynasty cast with inscriptions of "The 1st Year of Jian Wen" (1399)" and "Ying Tian Fu (Nanjing) Zao"

明代“万历元年”(1573)铁砣
An iron sliding weigh made in the first year of Wan Li (1573) in the Ming Dynasty

mentioned earlier, to weigh pieces of silver in purchasing and selling activities. Steelyards were also used as a common weighing apparatus. There is a hexahedral porcelain weight, on each side of which there is a Chinese character, and the six characters together read Gong Ping Jiao Yi (公平交易 Fairness in Trade), and Zheng Zhi (正直, Honesty). The knob of the weight is an irregular hexahedron, on two sides of which there are characters Jin Yu (金玉, Gold and Jade, meaning in Chinese "treasures") and Fu Gui (富贵, Richness and Honor) respectively. On the top of the porcelain weight there are characters explaining that it was made in the winter of 1619. Characters on the porcelain weight mentioned above were the reflection of capitalist ideas in business. They formed a sharp contrast with the inscriptions on measuring and weighing apparatus of the previous dynasties, which were characteristic of the state power. On some weights of the Ming Dynasty there were inscriptions about the calibrations done by stores and shops. With the development of the commodity economy, measuring and weighing apparatus, which were controlled by the ruling class and used for the collection of taxes, eventually became an instrument for fair trade.

Based on some of weights and some measuring apparatus available, it is calculated that 1 jin (斤) of the Ming Dynasty was equal to 596.8g, and 1 sheng (升) was equal to 1,035ml. Then it is clear that the measuring and weighing system in the Ming Dynasty followed the scales set in the previous dynasties.

明代“嘉靖九年”(1530)铁砣
An iron sliding weight made in the 9th year of Jia Jing (1530) in the Ming Dynasty

明代 铁权 铸“万历六年(1578)二月造”。重 5700 克
An iron weight, 5,700 g in weight, cast with inscriptions of "February of the 6th Year of Wan Li" (1578)

明代 瓷砣 书铭文“万历辛丑年”(1601)。重 598 克
A china sliding weight, 598 g in weight, with inscriptions in ink as "The Year of Xin Chou in Wan Li" (1601)

明代 瓷砣 书铭文“隆庆二年”(1568)。重 681 克
A china sliding weight, 681 g in weight, with inscriptions in ink as "The 2nd Year of Long Qing" (1568)

明代 瓷砣 书铭文“天平正合”
A china sliding weight with inscriptions in ink as “Tian Ping Zheng He”

明代 铜斗 铸阳文“成化兵子造”（1465～1487）和“福寿康宁”。容 9600 毫升
A bronze dou, 9,600 ml in capacity, in the Ming Dynasty with characters cut in relief as “Cheng Hua Bing Zi Zao” (1465-1487) and “Fu Shou Kang Ning” (Happiness, Longevity, Health, and Tranquility)

明代 瓷砣 书铭文“万历肆拾柒年孟冬月吉呈制”（1619）“公平交易正直”。重 571 克
A china sliding weight, 571 g in weight, with inscriptions in ink as “Made in Winter of the 47th Year of Wan Li” (1619)

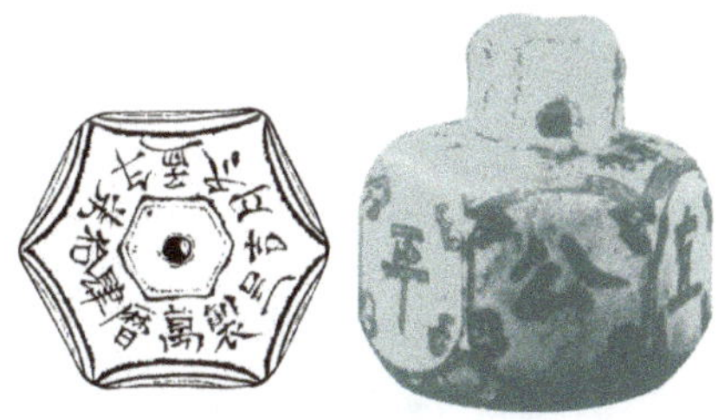

明代 戥秤 砣底部刻“万历年造”，最大秤量 20 两
Deng Cheng and its sliding weight, with inscriptions cast at the bottom as “Made in the Year of Wan Li”, for weighing objects of maximum 20 liang

明代 戥秤 砣底部刻“万历年造”。最大秤量 60 两
Deng Cheng and its sliding weight, with inscriptions cast at the bottom as “Made in the Year of Wan Li”, for weighing object of maximum 60 liang

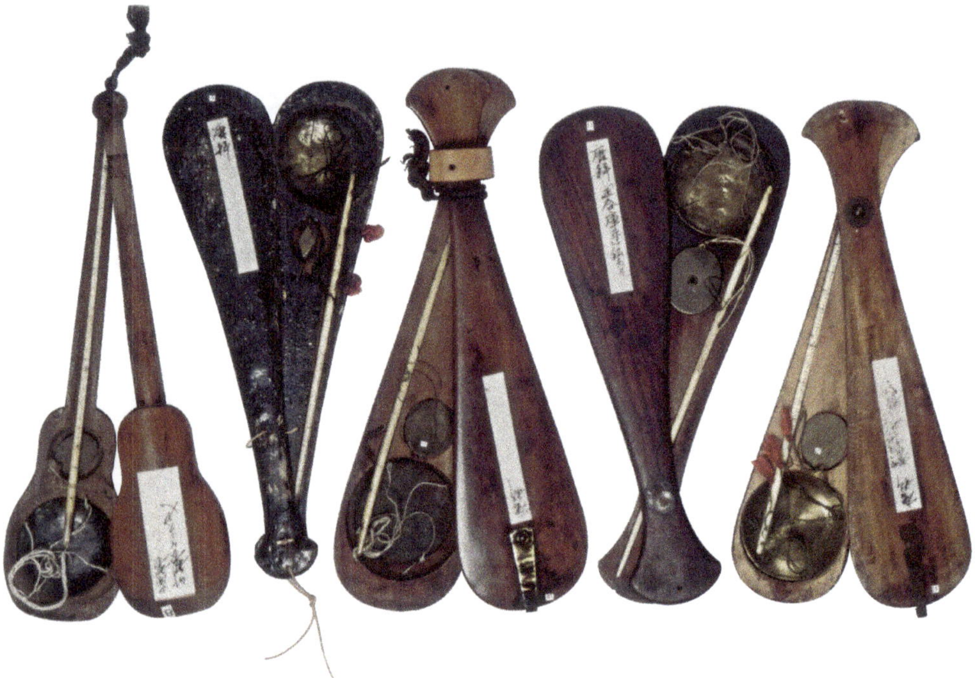

戥秤传入日本后被普遍使用，称“唐秤”

A Dengzi that became popular in Japan, known as "Tang Cheng" (Chinese Steelyard)

（2）清代度量衡标准的修订

清代度量衡制度始订于顺治（1644～1661），完成于康熙（1662～1722）、乾隆（1736～1795）。由于中国古代度量衡从一开始便与天文、乐律为一体，康熙、乾隆两位皇帝在研究天文、乐律时，也必然涉及度量衡。在康熙亲自主持下编定的《律吕正义》中说：横累100粒黍为一古尺，今则以纵累100粒为营造尺。黍是呈椭圆形的颗粒状物。为验证此说，我们曾实际做过校验：横排100粒约合23厘米，而纵排100粒约合32厘米。在另一本《数理精蕴》中，更专门列有度量衡表，说明用尺度来推导容量和重量，即：升，方三十一寸六百分；斗，方三百一十六寸；斛，方一千五百八十寸（明代以后五斗为一斛），名漕斛。经测算，1升合1035毫升。又以寸法定轻重之准，即：赤金每立方寸重十六两八钱，白银每立方寸重九两，红铜每立方寸重七两五钱，黑铅每立方寸重九两九钱三分，名库平。由营造尺、漕斛和库平两组成的度量衡制简称为营造库平制。

然而累黍尺毕竟复现性很差，金属的纯度又各异。这些不利因素都会影响其作为标准的精度。由于受当时科学技术条件的限制，实际上度量衡三者仍采用实物为最高标准。

乾隆对度量衡单位制也做了深入研究。乾隆七年（1742）亲自修订成书的《律吕正义后编》，再定度量衡制度。规定尺度和量器仍依康熙时所定制度，只是权衡仅以黄铜方寸重六两八钱为准，并由工部制造一批标准器发到全国各地。光绪三十四年（1908），政府又命农工商部及度支部拟定划一度量衡制度及推行章程：尺度仍依康熙纵黍尺之长定一尺，但因清营造尺之祖器（今称原器）遗失，便以《律吕正义》所绘营造尺图为依据复制成尺；量，仍以漕斛为准；衡，也仍以库平两为准，又因已认识到金属质地纯杂不尽相同，遂改用1立方米纯水的重量为重量单位。

清末尽管仍用旧制，却也与米制进行比较。今藏北京故宫博物院的一支铜尺和一件铜砝码，二器同装入一匣内，匣为朱红色，盖上有烫金楷书“高宗（即乾隆）纯皇帝钦定权度尺”和“户部库平、工部营造尺均遂旧制，与万国权度原器精校铸造”字样。尺一边刻营造尺10个寸格，另一边刻32厘米。砝码重37.3克（正合库平一两）。此尺与一两砝码正是清末重定度量衡制时，仍以康熙、乾隆确立的制度与米制进行比对的实物证据。清政府还向国际计量局定制了铂铱合金营造尺和库平砝码原器，作为营造库平制的最高标准。

清代 康熙皇帝考订古(秦汉)今(清)尺度,并亲躬累黍定尺,横排100粒(约23厘米)为古尺,竖排100粒(约32厘米)为清营造尺
Emperor Kang Xi, the Qing Dynasty, who was doing textual research: Collating and examining the values of lengths of rulers in the Han and Qin dynasties. He also personally participated in the setting of the length of the ruler by lining up a collection of 100 pieces of corn millet side by side (about 23 cm) as the length of the ancient ruler; and by arranging the 100 pieces of corn millet end to end (about 32 cm) as the length of Ying Zao Chi of Qing Dynasty

(2) The Improvement on Standards in the Qing Dynasty

The establishment of the measuring and weighing system in the Qing Dynasty started in the years of Shun Zhi (顺治, A.D.1644-1661), and completed in the regime of Kang Xi (康熙, A.D.1662-1722) and Qian Long (乾隆, A.D.1736-1795). Because the measuring and weighing system was related to astronomy and musical temperaments since ancient times, it was inevitable for the two emperors Kang Xi and Qian Long to associate their study on astronomy and musical temperaments with the measuring and weighing system. According to *Lv Lv Zheng Yi* (《律吕正义》) of which Kang Xi personally participated in the compilation, "an array of 100 pieces of corn millet arranged side by side was equal to the length of ancient chi (古尺). The length setting and production of ruler (尺) is based on the length of an array of 100 pieces of corn millet arranged end to end." This way of making rulers was called Ying Zao Chi (营造尺). Since the shape of a piece of corn millet is oval, we checked and found that the length was about 23 cm when 100 pieces of corn millet were arranged side by side; and about 32 cm when arranged end to end. In another book entitled *Shu Li Jing Yun* (《数理精蕴》), there was a table with a list of the units in the measuring and weighing system which described the ways of reaching the actual values of capacity and

清代各种形制的尺《律吕正义后编》列各种清尺的标准,分别为:营造尺长32厘米,裁衣尺长35.5厘米,律尺长25.9厘米
Variou types of rulers in the Qing Dynasty. In *Lv Lv Zheng Yi Hou Bian*, a collection of standard lengths of rulers was listed: Ying Zao Chi, 32 cm; Tailor's ruler, 35.5 cm; and Lu Chi, 25.9 cm

清代 从 1 分至 30 两铜砝码两套
Two sets of bronze weights in the Qing Dynasty: from 1 fen to 30 liang

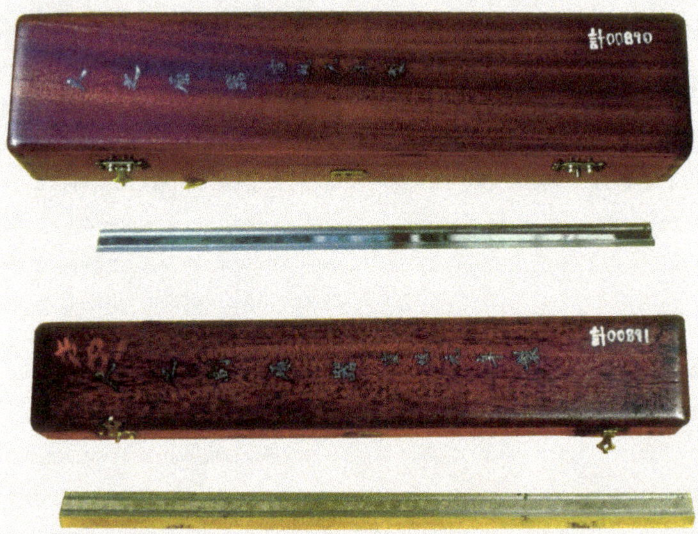

清代 宣统元年(1909)商请国际计量局制造两支营造尺:铂铱合金"尺之原器"和镍钢合金"尺之副原器"。长 32 厘米
The two Ying Zao Chi, 32 cm in length respectively, that were made as requested in the 1st year of Xuan Tong (1909) of the Qing Dynasty by the then International Bureau of Measures and Weights: "The Original and Standard Instrument for Ruler" made in platinoiridium, and "The Back-up Standard Instrument for Ruler" made in nickel steel

weights by applying for the lengths. Examples are shown below—1 sheng (升) = 31 cubic cun (寸)and 600 fen(分).1 dou (斗) = 316 cubic cun (寸).1 hu (斛) = 1,580 cubic cun (寸). After the Ming Dynasty, 1 hu (斛), known as cao hu (漕斛) was equal to 5 dou (斗).As was calculated, 1 sheng (升) = 1,035 ml. Cun (寸) was used as the base for measuring weights, namely, 1 cubic cun (寸)of gold = 16 liang (两) and 8 qian (钱), 1 cubic cun (寸) of silver = 9 liang (两),1 cubic cun (寸) of copper = 7 liang (两) and 5 qian (钱),1 cubic cun (寸)of lead = 9 liang (两), 9 qian (钱) and 3 fen (分).Such correlation between cun (寸)and the metals was named as Ku Ping (库平). In summary, the methods of making rulers for measuring lengths known as Ying Zao Chi (营造尺), setting the new units of capacity as cao hu (漕斛) and cubic cun (寸)for the units of weights was called Ying Zao Ku Ping System (营造库平制).

However, the precision of standards was affected by such factors as the complexity and reliability of setting the value of lengths by measuring a collection of corn millet (累黍尺), and the degrees of purity that varied among metals. With limitations of the science and technology at that time, the actual standards were adopted from real objects in setting the standard values of length, capacity and weight respectively.

Emperor Qian Long (乾隆) himself conducted in-depth researches on the unit setting for the measuring and weighing system. In 1742, the 7th year of Qian Long, the Emperor personally edited the book entitled *Lv Lv Zheng Yi Hou Bian* (《律吕正义后编》) that re-set the measuring and weighing system. While it was stipulated that the standard values for length and capacity remained the same as those set in the Kang Xi times, the new standard value of weight was set as—1 cubic cun(寸) of copper = 6 liang (两) and 8 qian (钱). The Emperor decreed to produce a set of new apparatus with the new standards to distribute all over China. In 1908, the 34th year of Guang Xu, the central government made a new decree on the Departments of Agriculture, Industry and Commerce, and the Division of Measurements to make plans and charters for unifying the measuring and weighing system, and to implement them. In the new Charter, the value of length (度) was following what had been set in the Kang Xi times, namely, the length of a collection of corn millet arranged end to end, but with the loss of the original mode in the early Qing times for making Ying Zao Chi (营造尺), the new ruler had to be re-produced according to the graph sketched in *Lv Lv Zheng*

清代 套装盒式 50 两砝码 总重 1831.7 克
A set of weights made in the Qing Dynasty, weighing 1,831.7g in total

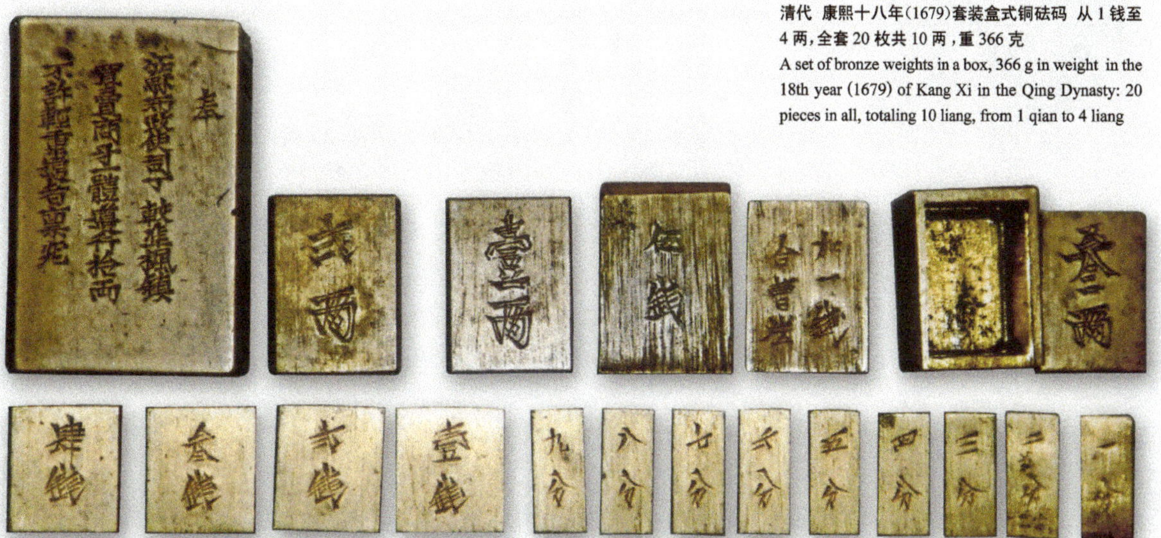

清代 康熙十八年(1679)套装盒式铜砝码 从 1 钱至 4 两,全套 20 枚共 10 两,重 366 克
A set of bronze weights in a box, 366 g in weight in the 18th year (1679) of Kang Xi in the Qing Dynasty: 20 pieces in all, totaling 10 liang, from 1 qian to 4 liang

Yi (《律吕正义》). Of the value of capacity (量), the liang (两) of Ku Ping (库平) was followed. With regard to weight, realizing that the purities varied among metals, the officials working on the new setting of correlations in Ku Ping (库平) changed to use cubes of pure water. Although the system followed the one set in the previous dynasty, comparisons were being made with the metric system (米制) during the last period of the Qing Dynasty. There exhibited in the Palace Museum in Beijing a bronze ruler (铜尺) and a bronze weight (铜砝码), both kept in a dark red box with characters saying that they were the Weight and Ruler (权度尺) inspected and confirmed personally by Gao Zong (高宗, namely, Emperor Qian Long). On the one side of the Ruler, there carved 10 scales for cun (寸格), and on the other side, there carved scales for 32 centimeters. The weight of the Weight was 37.3 g, just the same as 1 liang (两) in the Ku Ping (库平) system. These two objects have been evidence that comparisons were made between the metric system and the system set in the Kang Xi and Qian Long times towards the end of the Qing Dynasty when the measuring and weighing system was set again. The Qing Government even ordered a metal ruler from the then International Bureau of Measures and Weights (国际计量局) and two Ku Ping Weights (库平砝码) as the authoritative standards for making apparatus of measuring and weighing objects.

清代 乾隆三年(1738)"壹佰两"铜砝码

A bronze weight of "100 Liang" made in the 3rd year (1738) of Qian Long in the Qing Dynasty

清代 "嘉庆三年"(1798)"壹佰两"铜砝码

A bronze weight of "100 Liang" made in the 3rd year (1798) of Jia Qing in the Qing Dynasty

清代 康熙二十四年(1685) "五十两"铜砝码

A bronze weight "50 Liang " made in the 24th year of Kang Xi (1685) in the Qing Dynasty

清代 工部制造"五百两二两平" 铁砝码

An iron weight "502 Liang" prodnced by the Ministry of Works in the Qing Dynasty

清代 铜砝码 正面铸阳文"工部制造伍佰两正法马(砝码),会同户部较准颁发",腹围刻"乾隆二十九年十二月造"(1764)。重 18700 克

A bronze weight, 18, 700 g in weight, in the Qing Dynasty with characters cut in relief on the front as: "Standard Weight of 500 Liang, Poduced by the Ministry of Works, Inspected by and Promulgated with the Ministry of Revenue".Carved in the middle were the inscriptions of "Produced in December of the 29th Year of Qian Long" (1764)

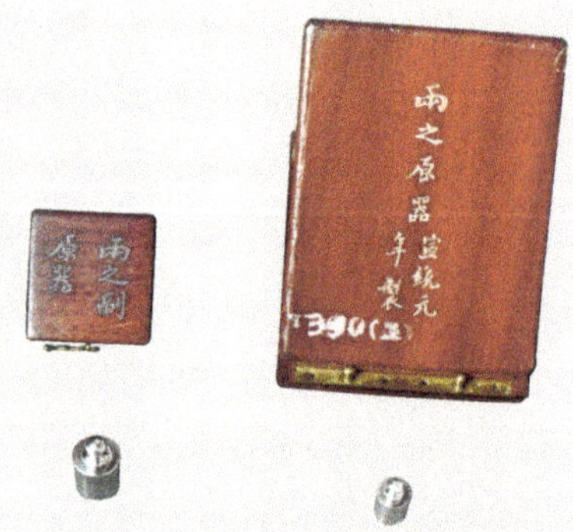

清代 宣统元年(1909)商请国际计量局制造库平两:铂铱合金"两之原器"和镍钢合金"两之副原器"。重 37.3 克

The two Ku Ping Liang, 37.3 g in weight respectively, that were made as requested in the 1st year of Xuan Tong(1909) of the Qing Dynasty by the then International Bureau of Measures and Weights: "The Original and Standard Instrument for Liang"made in platinoiridium, and "The Back-up Standard Instrument for Liang" made in nickel steel

清代 铜砝码 正面铸汉文"伍拾两,康熙二十年八月日造"(1681)。重 1862 克

A bronze weight, 1862 g in weight, in Qing Dynasty cast with inscriptions in Chinese on the front as: "50 Liang, Poduced in August of the 28th Year of Kang Xi" (1681)

清代 10 两至 200 两的标准铜砝码
A set of standard bronze weights, weighing from 10 to 200 liang, in the Qing Dynasty

200 两 200 liang | 100 两 100 liang | 50 两 50 liang | 30 两 30 liang | 20 两 20 liang | 10 两 10 liang

清代 天平及铜砝码
A Balancing Scale with a set of bronze weights in a box for balances, in the Qing Dynasty

清代 宣统元年(1909)商请国际计量局制造库平"五十两之原器"。重 1865.5 克
The Ku Ping Liang, 1,865.5 g in weight that was made, as requested in the 1st year of Xuan Tong (1909) of the Qing Dynasty by the then International Bureau of Measures and Weights: "The Original and Standard Instrument for 50 Liang"

清代 中国西藏地区称钱币专用 50 两铜砝码。重 1855.5 克
A 50-liang bronze weight, 1,855.5 g in weight, especially for weighing coins in the Tibetan region in China, in the Qing Dynasty

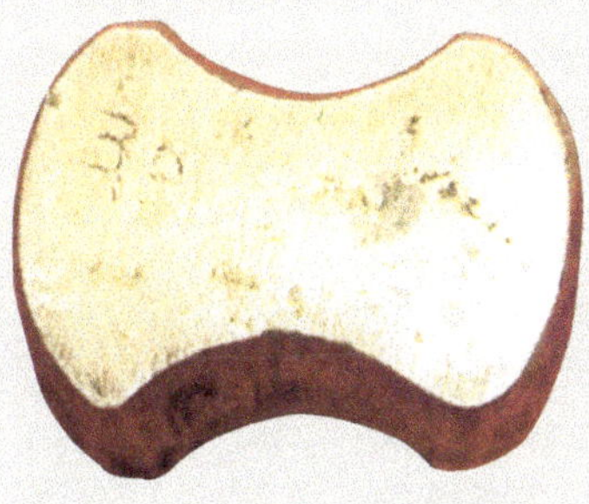

清代 中国西藏地区称钱币专用 30 两铜砝码。重 1118.6 克
A 30-liang bronze weight, 1,118.6 g in weight, for weighing coins in the Tibetan region in China, in the Qing Dynasty

清代 铜砝码 铭刻"叁拾两，三泉镇钱行公砝"。重 1102 克
A bronze weight, 1,102 g in weight, in the Qing Dynasty with inscriptions of "30 Liang, Standardized and Produced by the Qian Hang of San Quan County"

清代 铜砝码 铭刻 "叁拾两" "嘉庆丙子年(1816)置，嵩山会公议平，京货行"。重 1110 克
A bronze weight, 1,110 g in weight, in the Qing Dynasty with inscriptions of "30 Liang", "Equipped in the Year of Bing Zi (1816) of Jia Qing; Assessed by Songshan Hui; and Jing Huo Hang"

清代 铁砣 铸铭文"咸丰二年(1852)蔚县"
An iron sliding weight in the Qing Dynasty cast with inscriptions of "The 2nd Year of Xian Feng" (1852), Wei County"

清代 铜砣 铸铭文"辛未年造"(1871)"公平交易"
A bronze sliding weight in the Qing Dynasty cast with inscriptions of "Produced in the Year of Xin Wei" (1871) and "Fairness in Trade"

清代 铁砣 铸铭文"康熙四年"(1665)
An iron sliding weight in the Qing Dynasty cast with inscriptions of "The 4th Year of Kang Xi" (1665)

清代 铁砣 铸阳文"光绪十六年"(1890)。重 12300 克
An iron sliding weight, 12,300 g in weight, in the Qing Dynasty with inscriptions cut in relief as "The 16th Year of Guang Xu" (1890)

清代 石质砣 刻铭文"鱼行公秤足百斤""宣统二年"(1910)
A stone sliding weight in the Qing Dynasty with inscriptions of "Public Steelyard in Trade of Fish with Full 100 Jin" and "The 2nd Year of Xuan Tong" (1910)

清代 铜砣 铸铭文“清同治年”
A bronze weight made in the year of Tong Zhi in the Qing Dynasty

清代 铁砣 铸阳文“嘉庆十年(1805)十一月官秤” An iron sliding weight in the Qing Dynasty with inscriptions cut in relief as “Official Steelyard, November of the 10th Year of Jia Qing (1805)”

清代 铅质砣 铸铭文“康熙六年”(1667)“两广盐运”
Lead weights in the Qing Dynasty cast with “the 6th Year of Kang Xi ” (1667) and “for the transport of salt in Guangdong and Guangxi”

清代 铁砣 铸铭文“光绪颁发”“统税总局”
Iron weights in the Qing Dynasty cast with “Issued by Emperor Guang Xu” and “General Bureau of Tax Administration”

清代 一组兽形秤砣
A set of animal-shaped sliding weights in the Qing Dynasty

清代 度量衡检定牌 一面铸“雍正三年(1725)三月，府正堂”和“较准官秤，各行使用，倘有徇私，按法重惩”；另一面铸“押”
A Calibration Plate of Measuring and Weighing in the Qing Dynasty cast on one side with “March of the 3rd Year of Yong Zheng (1725)” and “The Standard Steelyard for All Firms; Severe Punishment by Law on Violations”. Cast on the other side was a character “Ya” (Station)

清代 “道光”“光绪”年间的瓷砣
China sliding weights in the Qing Dynasty popular during the years of “Dao Guang” and “Guang Xu” times

清代 铁方升 铸阳文“户部样仓升，康熙五十四年十月造”(1715)。容 1043 毫升
An iron Fang Sheng, 1,043 ml in capacity, with inscriptions cut in relief as “Standard Cang Sheng (weighing instrument) produced in October of the 54th year of Kang Xi (1715) and promulgated by the Ministry of Revenue”

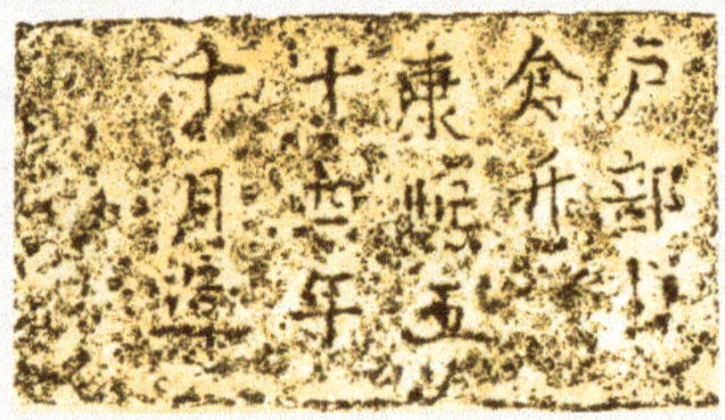

清代 木斛 书写“邑庙公斛，奉上海县正堂，嘉庆十八年(1813)较准”。容 50294.5 毫升
A wooden hu, 50,294.5 ml in capacity, with characters of “Standard and Public Hu in Yi Miao Produced in Agreement with the Standard Promulgated by the Shanghai County in the 18th Year of Jia Qing (1813)”

清代 铁斛 铸阳文"乾隆二十三年(1758)遵部颁式样造铁斛"。容 56000 毫升
An iron hu, 56,000 ml in capacity, with inscriptions cut in relief as "Standard Iron Hu Produced in Agreement with the Standard Promulgated by the Ministry of Revenue in the 13th Year of Qian Long (1758)"

清代 "大清宣统元年"(1909) 木斗
A wooden dou made in the first year of Xuan Tong in the Qing Dynasty (1909)

清代 "光绪十二年"(1886)木斗
A wooden dou made in the 12th year of Guang Xu (1886)

清代 "苏州府"木斗
A wooden dou made by "Suzhou Fu" in the Qing Dynasty

清代 鎏金铜嘉量 乾隆亲定度量衡时，以新莽嘉量为准则，研究秦汉至明清度量衡沿革而敕制。今仍陈列在北京故宫博物院太和殿前
A gold-gilded bronze jia liang in the Qing Dynasty. Emperor Qian Long, when examining and formulating the measuring and weighing systems with Xin Mang Jia Liang as the standard, explored the evolution of the measuring and weighing systems from the times of Qin and Han to those of Ming and Qing, and issued an imperial edict for producing this weighing instrument, which is still displayed in front of the Tai He Hall in the Palace Museum in Beijing

THE HISTORY

OF ANCIENT CHINES

MEASURES AND WEIGHTS

11

THE HISTORY OF ANCIENT CHINESE MEASURES AND WEIGHTS

中国古代计量史

十一、元明清时期的计量技术

Chapter Eleven

Measuring Technologies in the Yuan, Ming and Qing Dynasties

1. 郭守敬创高表测影

郭守敬是元代杰出的科学家，他具有多种学识和才能：在仪器制造、地形测量、农田灌溉、河道修建等领域都取得了卓越成就。他在天文仪器制造方面有惊人的创造力，曾先后制作了简仪、仰仪、高表、候极仪、浑天象、景（影）符、方日晷等 12 种新仪器。对传统立表测影技术的改进是他诸多成就之一。用圭表测影，表高 8 尺。夏至日阳城（今河南登封告成镇）正午日影长 1 尺 5 寸是西周以来的古制，至元代历时已 2000 多年，几乎无人敢轻易作变更。为了提高测量日影的精确度，郭守敬大胆创新，勇于实践。在古人测影的阳城建立了一座测影台，立高表 40 尺，是传统表高的 5 倍，这样所测日影长度的相对误差便可以减少到 1/5。为了取影方便，又在表的顶部架了一根横梁。按常理加长表高就能使日影变化更显著，但是表增高后，照射在圭面上的日影又会虚而淡。为了克服高表影虚的缺陷，郭守敬设计了一件附属仪器——景（影）符。景（影）符是根据针孔成像原理制成的，即在一个小框架上安一个可以转动的铜片，中间开一个小孔。当太阳过子午线时，把景（影）符放直在水平的圭面上，南北移动并转动铜片，使照射在圭表顶部横梁上的日光首先通过景（影）符上的小孔，再照射到圭面上，形成一个米粒大小，中间含有铜横梁的太阳倒像。当横梁影正好平分日像时就得到了日面中心影长。利用影符基本解决了由于高表造成“影虚而淡”的问题，极大地发挥了高表测影的优越性，从而达到提高测影准确度的目的。

郭守敬建造的测影台，台前有沿着地面向北延伸的一条石圭，用 36 块方石拼接而成。圭面上有两条平行的水槽，起着水平作用，两槽中间狭长的圭面即是量度日影的圭表尺，亦称“量天尺”。尺全长 3129.4 厘米，按《元史》所记，圭长 128 尺，得元代量天尺为 24.45 厘米。可见郭守敬造天文仪器的专用尺仍为隋唐以来的“小尺”。

元代科学家郭守敬
Guo Shoujing, a scientist in the Yuan Dynasty

元初郭守敬设计的登封测影台，其表高从传统的 8 尺提高至 40 尺。圭上所刻量天尺一尺长 24.5 厘米
The Sundial Platform in Dengfeng (County) designed by Guo Shoujing during the early years of the Yuan Dynasty, whose gnomon was heightened from the previous 8 chi to 40 chi. Carved on the plate was the length of 1 chi for Liang Tian Chi (24.5 cm)

郭守敬设计的方晷，根据测定的日影方向和长度来定方位角度。可与漏刻配合使用
In determining the positions and angles, Guo Shoujing invented a Square Sundial to measure the direction of the shadow and its length. This kind of sundial could be used together with a Lou Ke (Clepsydras)

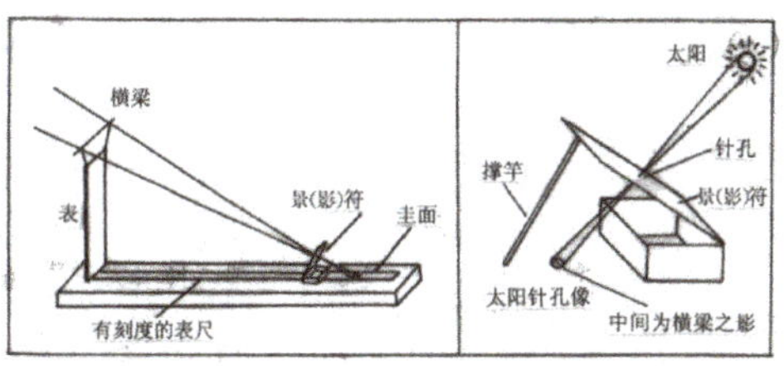

为解决表高而影虚的问题，郭守敬根据针孔成像原理设计了景(影)符。左图：针孔成像示意图。右图：调节景(影)符倾斜度，使之垂直于日光
In solving the problem of diffusion of the shadow projected by the gnomon when it was heightened, Guo Shoujing invented Ying Fu, a device which was on the basis of the principle that images could be produced through a narrow aperture. On the left: A diagram of images produced through a narrow aperture. On the right: Adjusting the tilted Ying Fu so that it was perpendicular to the sunlight

郭守敬用针孔成像原理设计制造的仰仪
A diagram of an instrument (Yang Yi) invented by Guo Shoujing in the light of the principle that images were produced through a narrow aperture

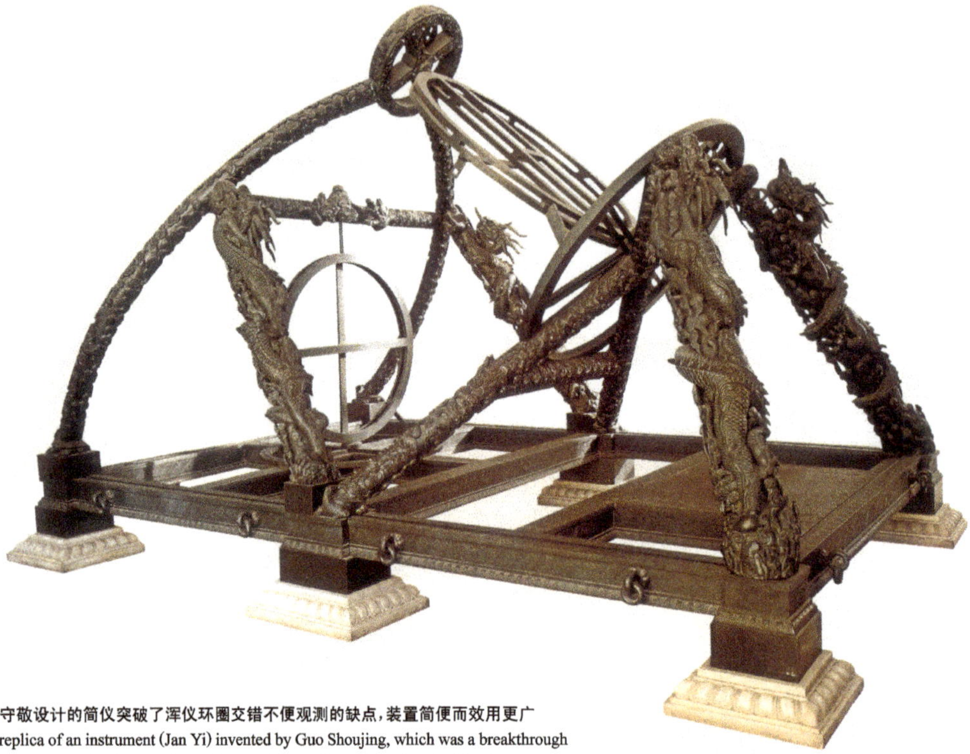

郭守敬设计的简仪突破了浑仪环圈交错不便观测的缺点，装置简便而效用更广
A replica of an instrument (Jan Yi) invented by Guo Shoujing, which was a breakthrough in overcoming the shortcomings of an armillary sphere whose interlocking hoops made observation difficult. Hence this device was simpler and more effective in application

1. Gao Biao (高表，Sundial) Invented by Guo Shoujing (郭守敬)

Guo Shoujing (郭守敬), a renowned scientist in the Yuan Dynasty, was endowed with versatile and brilliant talents. He accomplished splendid achievements that were embodied in device making, geological measuring, and irrigation in agriculture and other areas. His creativity was demonstrated in astronomical devices, such as Jian Yi (简仪), Yang Yi (仰仪), Gao Biao (高表), Hou Ji Yi (候极仪), Hun Tian Xiang (浑天象), Ying Fu [景(影)符], Fang Ri Gui (方日晷，Sundial) and so on, totaling 12 devices new at that time.To innovate on the traditional method of erecting rods for measuring the lengths of the shadow was one of his achievements.

There was a Gui Biao (圭表，Sundial) with a height of eight chi (尺) standing in Yangcheng [阳城，now called Gaocheng (告成), a town located in Dengfeng(登封), Henan Province]. On Xia Zhi (夏至，Summer Solstice), the longest day in the year when the sun was at its highest position at noon, the length of the shadow used to be measured as 1 chi (尺) and 5 cun(寸), in the ancient measuring system ever since the Western Zhou Dynasty (西周). Up to the Yuan Dynasty (元代), no one had dared to question or even adjust this observation. For the purpose of raising the precision in measuring the length of the shadow, Guo Shoujing took the initiative to innovate and put his efforts in practice. He erected a platform in the town Yangcheng (阳城) where ancient people measured the length of the shadow, with a rod of 40 chi (尺) long, which was 5 times the length of the traditional one. In this way, the discrepancy as a result of measurement would decrease to 1/5. In order to measure the length of the shadow in a handy way, he laid a beam horizontally on the top of the rod. A common sense was that the changes of shadow would be more obvious with the length of the rod increased. However, because of such increase, the shadow over the face of Gui Biao would be vague or even blurry. Guo Shoujing invented an attached device called Ying Fu to overcome the shortcomings caused by the greater length of the rod. This attachment mainly comprised a piece of coin-shaped flat copper with a tiny hole at the center. The

明代 正统四年(1439)造浑天仪 今陈列在南京紫金山天文台
An armillary sphere made in the 4th year of Zheng Tong (1439) in the Ming Dynasty which is still displayed at the Nanjing Zijinshan Astronomical Observatory

principle of adopting such a device was that when hung over the face of Gui Biao and moved to and fro northward and southward, the hole would catch instantly the sun light the moment it was over the top of the rod. When the horizontal beam divided the sun evenly, the reflection of the sun was caught through the hole with a size as small as that of a piece of millet. The advantages of such an invention were conspicuous. It not only overcame the shortcomings of a longer rod with a vaguer shadow, but also increased the precision in measuring the actual length of the shadow.

In front of Gui Biao platform, there was a flat Stone Plate (石圭) expanding itself northwards, which consisted of 36 pieces of square stones. On the flat surface of Stone Plate, there were two parallel gutters functioning to ensure that the surface was leveled. The narrow part of the surface between the parallel gutters was known as Gui Biao Chi (圭表尺), also called Liang Tian Chi (量天尺), with a length of 3,129.4cm. According to what was recorded in *Yuan Shi* (《元史》), the length of the Stone Gui was 128 chi (尺). Liang Tian Chi was 24.45 cm in length. This indicated that the special ruler Guo Shoujing used for constructing astronomical devices was following Xiao Chi (小尺, shorter ruler) since the Sui and Tang dynasties (隋代, 唐代).

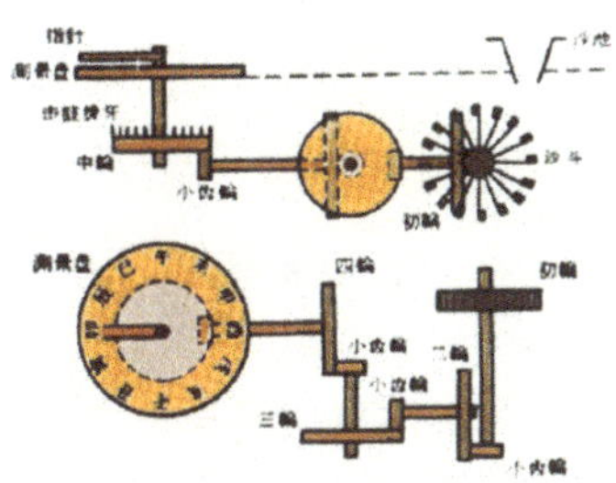

沙漏结构图
An illustration of the mechanics of a sand clock

明代 沙漏示意图
A diagram of a sand clock in the Ming Dynasty

明代 铜圭表 圭上所刻量天尺长 24.5 厘米
A bronze sundial in the Ming Dynasty with the length of Liang Tian Chi carved on the plate as 24.5 cm

元代 延祐三年(1316)造四级式漏刻，由日壶、月壶、星壶和受水壶组成

A 4-grade clepsydras made in the 3rd year of Yan You (1316) in the Yuan Dynasty that comprised 4 containers: the Sun Vessel, the Moon Vessel, the Star Vessel, and the Water Vessel (to receive and hold the water dripping in)

2. 元明时期的计时仪器

郭守敬制造了不少计时机械，其中大明殿灯漏是十分突出的一项。灯漏是一个外形像灯笼球，用水力推动的机械报时器，上面布置有能按时跳跃的动物模型，同欧洲在机械钟表上附加种种表演机械性质相同。灯漏分为机械传动结构和灯球两部分。整个计时器的动力和齿轮系等构件全部“隐于柜中，以水激之”。动力机构与水运浑象仪相似。灯球部分较为复杂，共分四层。虽然没有演示星空运行的内容，但在第一层有日、月、参、辰。第二层为“龙虎鸟龟之象”。每至古一刻动物模型分别跳跃，同时还以音响报时。第三、四层也是报时的机构，每至一个时辰就有手拿相应报时牌的木人出来通报。其形式也与水运浑象仪类似，但配以不同音响，即采用分别敲击钟、鼓、钲、铙四种乐器来报告各个时刻的方法，比以往的报时器更为巧妙，更为完美。

明代计时器中值得一提的是沙漏。据《明史》记载：“明初，詹希元以水漏至严寒水冻辄不能行，故以沙代水。”沙漏由沙池、斗轮(初轮)、减速齿轮和时刻显示四部分组成。沙池底部有一个小孔，用来贮放和漏出细沙。斗轮的周围均匀地分布着 16 个沙斗，当沙池中漏出的细沙注入其中一个沙斗时，便会导致斗轮重心倾斜而转动，从而带动四对分别为 6 齿和 36 齿组成的减速齿轮组，第 4 个齿轮(中轮)的轴则穿出一只盘面刻有 12 时辰和昼夜百刻标志的测时盘盘心，立轴上端装有指针。沙漏运行时，指针也就相对于测时盘而旋转。从沙池漏出的沙流推动斗轮经过齿轮减速组，使最后一只齿轮带动的指针在一天中恰好旋转一周。沙漏的轮与沙池皆藏在机腹，唯盘露出机面，两旁各立一黄衣童子：一击鼓，一鸣钲。形式上更接近现代的机械钟了。

明代以后，具有中国民族特色的机械钟表就很少出现了。17 世纪西方资本主义经济得到迅速发展，其机械钟表制造工艺也日趋成熟。随着欧洲传教士的到来，西方机械钟表也进入中国。

2. Time Reckoning Instruments in the Yuan and Ming Dynasties

The scientist Guo Shoujing invented a good number of time-reckoning devices, among which Da Ming Dian Deng Lou (大明殿灯漏) was outstanding. Deng Lou (灯漏) was a lantern-shaped hydraulic device with animal models that would jump according to the time elapsed. This was similar to the mechanical clocks made in Europe. Deng Lou consisted of a mechanical structure and a lantern ball (灯球). Water provided the device with the source of power to move the internal parts including the gears, which was similar to those of Shui Yun Hun Xiang Yi (水运浑象仪). The part of the Lantern Ball was complex with four layers. At the first layer, there were orbits of the sun, the moon, and so on. There were animal models at the second layer including Dragon, Tiger, Bird and Tortoise that would jump in turn as the time went on. They even signaled the time with different kinds of sound. At the third and fourth layers, there were also mechanisms for telling the time. This was carried out by a wooden puppet holding a pertinent piece of board with the time indicating on it. Such mechanism was different from Shui Yun Hun Xiang Yi (水运浑象仪) in a way that it reported the time with various kinds of sound made by knocking

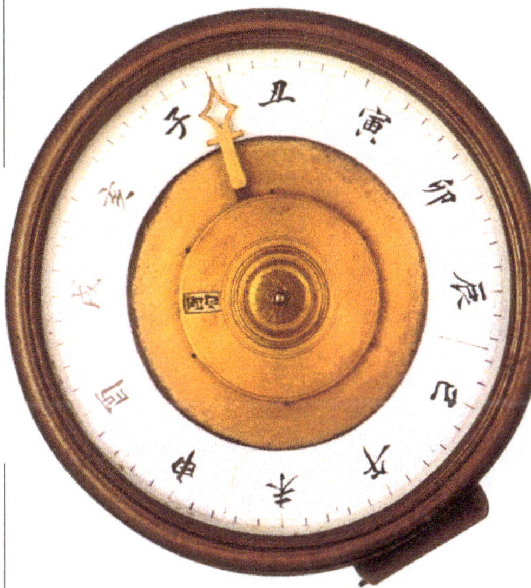

清　时辰醒钟　由宫廷钟表处制造，以中国传统的一日十二个时辰为表盘显示

A shichen alarm clock in the Qing Dynasty made in the workshop in the Imperial Court, whose face showed 12-hour periods, the traditional Chinese method of time reckoning (there are 12 shichen in a day)

图间插文 Descriptions matching the Pictures

明代后期，欧洲传教士来到中国，也带来了令朝廷内外为之着迷的自鸣钟。明末清初，民间已有人相继学制机械钟。康熙十六年（1677）前后，清宫内设置了钟表作坊，至康熙三十年（1691），因规模扩大、工匠增多而迁至宫外，另设钟表作坊。此外在广州、苏州、南京等地也先后开设了钟表作坊。

During the last years of the Ming Dynasty, European missionaries came to China, bringing with them chime clocks that fascinated those staying inside the imperial court and people in the street alike. At the end of the Ming Dynasty and the beginning of the Qing Dynasty, some craftsmen followed suit to make such mechanical clocks. During the 16th year of Kang Xi (1677), workshops were set up inside the Palace to make these kinds of clocks. In the 30th year of Kang Xi (1691), the workshops were moved outside the Palace owing to an expansion of the production scale and an increase of the number of craftsmen. More workshops were also opened in Guangzhou, Suzhou, Nanjing and other places.

清代　康熙十二年（1673），按比利时传教士南怀仁所绘图样制造的赤道经纬仪。主要用来测定太阳时、天体的赤经差和赤纬。从总体上看，仍然处在中国传统天文学框架中

The Globe with the Celestial Equator made in the 12th year of Kang Xi (1673) in the Qing Dynasty, which was on the basis of the drawing by Ferdinand Verbiest, a missionary from Belgium.It was mainly used for tracing the sun, the deviations of the celestial equators and latitudes of heavenly bodies. Generally speaking, the technogy was still in the developmental stage of Chinese traditional astronomy

on the four types of musical instruments of bell, drum, zheng (钲), and nao (铙). This was more interesting and wonderful.

Another time-reckoning device in the Ming Dynasty worth mentioning here was Sha Lou (沙漏). As was recorded in *Ming Shi* (《明史》, *the Codes of Ming*), in the early period of the Ming Dynasty, Zhan Xiyuan (詹希元) noticed that Shui Lou (水漏, a device using water) was not functioning in winter when the water was frozen. He replaced Shui Lou with a device using sand (沙) and hence it was called Sha Lou (沙漏). It consisted of four parts, namely, a Sha Chi (沙池 a platform to hold the sand), a Dou Lun (斗轮, a wheel with sand containers hanging around it), a breaking gear and compass with scales of time. There was a small hole at the bottom of the Sha Chi to let the sand go through. There were 16 sand containers called Sha Dou (沙斗) scattered evenly around Dou Lun. When one of Sha Dou received the sand going through the hole, it would tilt because it was getting heavier, thus causing Dou Lun to move, which in turn moved the four pairs of gears and breaking gears to turn. Attached to the central gear was the center of the compass with scales of time, indicating 12 shi chen (时辰, about 2 hours). A needle was connected to the axle at the center of the compass. The moving gears made the needle go around the face of the compass. Such mechanism was accurate enough to make the needle turn one round in one day and night. While the wheel and sand platform were hidden, with the compass face sticking out, it looked like a modern clock. On each of the two sides of the compass surface stood a puppet boy in yellow clothes. One would knock at the drum and the other at zheng (钲).

Since the Ming Dynasty, there were fewer and fewer such kind of mechanical clocks made in China with nationality features. As the West was developing rapidly in its capitalist economy from the 17th century, the technologies in producing mechanical clocks were becoming mature. These types of clocks were brought into China when European missionaries came to China.

清代 镀金转花钟 由广州制造，表盘采用罗马数字显示，一日转两圈，共24小时
A gold-plating clock in the Qing Dynasty made in Guangzhou, with Roman numerals on the face. The hour hand goes around the clock twice in 24 hours

3. 物质密度的测量与在度量衡标准上的应用

中国古代很早就对物质的密度有所认识。《孟子》曰："金重于羽者，岂谓一钩金与一舆羽之谓哉。"注曰：金比羽，"谓多少同而金重耳"。早在战国时期人们已认识到物质密度是指单位体积的物质质量。对物质密度的测定见于《汉书・食货志》："黄金方寸，而重一斤。"《后汉书・礼仪志》："水一升，冬重十三两。"最晚至汉代，人们对密度的认识已从定性走向定量。历代对固体物质的密度做过测定，还见于《九章算术》《孙子算经》等。清初汇编的《数理精蕴》曾以较精密的方法测定了金、银、铜、铁、锌等32种物质的密度。并指出："凡物，知其体积即知其重轻，知其重轻即知其体积，而权度无遁情也。"

度量衡的实物标准，实际上也是利用密度的原理来制定的。如国际标准公斤(千克)砝码"公斤(千克)原器"，是一个直径和高都是39毫米的圆柱体，用90%的铂金和10%的铱金制成。米制又将长度、体积和质量的量度相互关联起来：边长为10厘米的立方体体积为1000立方厘米，即1升。1升水在4摄氏度下质量为1公斤(千克)。汉代虽然已曾有黄金方寸重1斤，水1升冬重13两的记述，却受当时科学技术条件的限制，并未真正运用到实际中去。到了南北朝，北魏道士李兰创制秤漏时，规定："漏水一升，称重一斤，时经一刻。"应该是指一升水的质量正合一斤，此与北朝一升约合600毫升，一斤约660克约略相合。又以所称一斤重之水作为衡量计时器准确性的一个标准。到了北宋，李照创制水秤，他的设计思想就是以一升水之重来定一斤的标准，可惜未被官方采用。但他这一理论上的研究和实践，在古代度量衡史上却是值得称颂的。到了清代，康熙御定《数理精蕴》列度量衡表，皆以营造尺之寸法定砝码之轻重，即以赤金、白银、红铜、黑铅各铸为方寸，再实测各种金属的密度。至乾隆御订《律吕正义后编》，又规定唯以"黄铜方寸重六两八钱"作为权衡的标准。然而皆因金属纯度各异，使这种标准的精度受到影响，遂改"以营造尺一立方寸纯水量之重为权之重率。"可见以物质密度定度量衡标准的科学性早已被认识，2000多年来一直在不断地被探求和改进。

《后汉书・礼仪志》记："水一升，冬重十三两"
In *Hou Han Shu / Li Yi Zhi*, it was stated that "1 sheng of water weighed 13 liang in winter"

3. Density Measuring and Its Application in the Standard Setting of Measures and Weights

People in ancient times had the knowledge of density. In *Men Zi* (《孟子》) by Mencius (孟子), it was described that "as gold is much heavier than feather, how can they be in comparison with each other?" and "How many pieces of feather would it take to reach the same weight as a piece of gold?" As early as in the Warring States(战国), people realized the concept of density as the quantity of weight in relation with its unit of volume. An early description of such measurement was seen in *Han Shu / Shi Huo Zhi* (《汉书・食货志》) that read, "the weight of a cubic cun (寸) of gold is 1 jin (斤)" In *Hou Han Shu / Li Yi Zhi* (《后汉书・礼仪志》) it read, "the weight of 1 sheng (升) of water is 13 liang (两)". Till the Han Dynasty, people developed the concept into quantitative description. Also there were experimentations undertaken on solid matters in the past dynasties. In *Jiu Zhang Suan Shu* (《九章算术》) and *Sun Zi Suan Jing* (《孙子算经》), those activities were recorded in details. In *Shu Li Jing Yun* (《数理精蕴》) compiled in the early period of the Qing Dynasty, records showed the methods used for examining the density of 32 matters like gold (金), silver (银), copper (铜), iron (铁), zinc (锌) and so forth. It was also quoted as saying that "Of a matter, to tell the weight is to tell the volume, regardless of the instruments for weighing or measuring (权度) it".

As a matter of fact, the setting of standards of objects determined to use in the measuring and weighing system is an application of such principle of density. Take the "Standard Kilogram" (千克原器), the international standard weight for 1 kilogram (国际标准千克砝码), for example. It is a column with both the diameter and height as 39mm, containing compound metals as 90% of platinum and 10% of iridium. In the metric system (米制), the measurements of length, volume and weight are associated with each other. The volume of an object with 10cm in length, width and height is 1,000 cubic centimeters, that is, 1 sheng (升). One sheng (升) of water, at the temperature of 4℃ weighs 1kg. Although people in the Han Dynasty (汉代) acquired the knowledge that 1 cubic cun (寸) of gold weighed 1 jin (斤); and 1 sheng (升) of water weighed 13 liang (两) in winter, such concept was not applicable, owing to the limitations of science and technology at that time. It was not until the times of the Northern and Southern Dynasty (南北朝) when a Taoist named Li Lan (李兰) living in Northern Wei (北魏) linked the concept to practice. At the time of inventing the Cheng Lou (秤漏), he set clearly the associations among the amount of water, unit in weight and scale in time, as "1 ke (刻 15 minutes) has elapsed when 1 sheng (升) of water, weighing 1 jin (斤), goes through the device". It should be pointed out that the key idea in Li Lan's description was that the weight of 1 sheng (升) of water was 1 jin (斤), which was coinciding with what was described in the Northern Dynasty (北朝) as "1 sheng (升) was approximately equal to 600ml" and "1 jin (斤) was about 660g". Also the amount of water weighing 1 jin (斤) was set as the standard for examining the precision of time

《汉书・食货志》记："黄金方寸，而重一斤"
In *Han Shu / Shi Huo Zhi*, it was stated that "the weight of 1 cubic cun of gold was 1 jin"

reckoning devices. In the Northern Song (北宋) times, Li Zhao (李照) invented Shui Cheng (水秤, a hydrometer), the design of which reflected the standard of 1 sheng (升) of water weighing 1 jin (斤). It was a pity that his invention was not adopted officially. But he was credited for applying the theoretical ideas to practice, for the first time in the history of the development of the measuring and weighing system.

In the Qing Dynasty, Emperor Kang Xi (康熙) decreed to write up *Shu Li Jing Yun* (《数理精蕴》) with lists of units in the measuring and weighing system. The standards reflected in the lists, in terms of length, weight and density, were the officially confirmed weights (法定砝码), which were made according to the value of cun (寸) set in Ying Zao Chi (营造尺), namely, the weights were made in gold, silver, copper, and lead in cubic cun (寸), and density was determined by actually weighing the respective object. Later on Qian Long (乾隆) decreed to compile *Lv Lv Zheng Yi Hou Bian* (《律吕正义后编》) with the stipulation as that "the standard set for weighing is 1 cubic cun (寸) of copper that weights 6 liang (两) and 8 qian (钱)". However, because the accuracy was affected owing to the variations of purity in different metals, later on the standard for weighing was re-set according to the weight of fresh water of 1 cubic cun (寸) of Ying Zao Chi (营造尺). These efforts manifested that the science of relating the density of a substance to the standards in the measuring and weighing system was acquired by people as early as in the ancient times, and that this science has been explored and developed for more than two thousand years continuously.

国际"公斤(千克)原器" 米制公约规定，1 升水在 4 摄氏度下质量为 1 千克

The International "Standard Kilogram". According to the regulations set in the Metric System Treaty, the mass of 1 liter of water is 1,000 g at 4 ℃

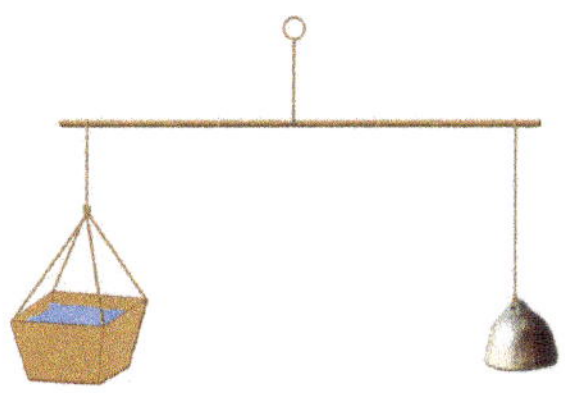

清代 康熙御定《数理精蕴》列度量衡表。为准轻重之率，以赤金、白银、黑铅等各铸为立方寸，再测各种金属的密度

Emperor Kang Xi issued his imperial edict to compile *Shu Li Jing Yun* that listed tables of the measuring and weighing system. In order to set the standards for weight, gold, silver and lead were cast in cubic cun respectively, and the specific density of each was then determined

北宋时期 李照设计制造了一种水秤，提出用一升水的质量来定一斤的标准，即"一合之水重一两，一升之水重为一斤"

In the Northern Song Dynasty, Li Zhao designed and invented hydrostatic balance, a kind of water scale, and advocated a standard of taking the specific weight of 1 sheng of water as 1 jin, namely, "The weight of 1 ge of water was 1 liang, and 1 sheng of water was 1 jin"

4. 明清时期的测量技术

明代不仅国内经济繁荣，而且海外贸易也很繁荣。在永乐三年至宣德八年(1405 ～ 1433)近 30 年间，郑和“七下西洋”，到达亚洲、非洲 30 多个国家和地区。在繁荣对外贸易的同时，也促进了手工业的发展。特别是造船业，其规模之大、制造工艺之先进，都远非宋、元两代可比。据《明史 · 郑和传》记载：郑和出使西洋的海船，最大的长 44 丈，广 18 丈，可乘载 500 至 1000 人。连续七次的出海航行，使郑和与其同僚们掌握了大量的航海知识，并完成了举世闻名的《郑和航海图》，其中也包括了对航海里程、海水深度等测量技术的记述。

当时记录航海里程的单位是“更”，大约船行一昼夜的里程为 10 更，每更行程约为 60 里。但受海流、风向及风速的影响，一更也可以大于或小于 60 里。他们总结出一些简单易行的检测方法：以木片于船首投入海中，人从船首速行至船尾，木片与人齐至，则更准；若木片比人先至，则为过更；若人行至船尾，而木片未至，则为不上更。明代航海者还根据前人的经验，利用测量海水的深浅来辅助判断船行的地理位置。其测水深的单位为“庹”，一庹等于成人两臂平伸的长度，约 160 厘米。测水深的办法是，用长绳系一铅锤等重物，沉入海中，再量绳长。锤底涂以牛油，借其黏附物探明海床是泥沙还是岩石，然后结合前人记载及航海的经验，判明船所在的位置。

17 世纪中叶，西方传教士和商人已踏入中国，此时也正是清朝国内统一、经济繁荣的鼎盛时期。康熙很注意对天文历法的改进。他发现西方的历法更科学，于是颁行比利时传教士南

宋代 手持罗盘的陶俑
A terra-cotta figurine holding a comapss, in the Song Dynasty

宋代 水浮式指南针 在航海中得到广泛使用
A compass (with a floating needle) in the Song Dynasty, widely applied in navigation

明代 郑和宝船复原图 明代造船业已十分发达，为郑和七下西洋做好了物质保证

A diagram of the restored Zheng He Ship in the Ming Dynasty, the ship building industry was developed, which was a substantial assurance for the Seven Voyages by Zheng He's Fleet to the "Western Ocean"

怀仁制定的《时宪历》，并且命南怀仁按欧洲的先进方法结合清朝度量衡制度督造天文仪器。这期间，在地图测绘方面也取得了显著的成就。中国古代地图测绘虽然有悠久的历史和杰出的成就，但毕竟受当时的科技水平所限，里程计算不够精确。康熙四十七年(1708)朝廷决定在全国范围内进行一次大规模的地图测绘工作，采用了西方先进的测绘技术和地理知识。为了保证测量的准确一致，首先统一了测量中使用的长度单位，规定以工部营造尺为标准尺度，五尺为一步，360 步为一里。规定纬度一度为 200 里。这种以地球形体来规定尺度的方法，也是世界上最早的。前后用了 10 年时间将测量结果汇总后，绘制成全国地图《皇舆全览图》，达到了地图绘制史上极高的水平。李约瑟曾说："它不但是亚洲地图中最好的，而且比当时欧洲地图更精确。"

康雍乾时期对农业生产都很重视。在此 100 多年间，无论是治理河道还是水利的开发，都取得了可观的成就。在这方面有突出贡献的人是陈潢。陈潢特别注意因势利导地疏通水流，为了查明黄河水情、地势，他"必亲历度势为第一事"。沿河"跋涉险阻，上下数百里，一一审度"。甚至在疾风中驾轻舟，测水之深浅。为了准确地控制洪水，陈潢"以推测土方之法，移而推测水方"。创造了"测水法"，"以水纵横一丈、高一丈为一方"。把河水的横切面积乘以流速来计算水流量。他说，河道闸口水的流速和一昼夜的流量，都可以通过测量和计算得到。"其法，先量闸口阔狭，计一秒所流几何，积至一昼夜，则所流多寡可以数计矣。"陈潢在水利工程方面取得的成就与他创立的测水法是分不开的。

明代 郑和七下西洋路线图

A navigation map for the Seven Voyages by Zheng He's Fleet to the "Western Ocean", in the Ming Dynasty

明代 《郑和航海图》中有对航海里程、海水深度等测量的记述。测量水深浅的单位是"庹"，测量方法是用长绳系一铅锤沉入海中，再测绳长

In the book entitled *Zheng He Navigation Maps*, there were records of measuring the mileage, depth of the seas, etc. The unit of length for measuring depth was tuo, and the method for determining the depth was to sink a rope with a lead plummet and then to measure the length of the rope

4. Measuring Technologies in the Ming and Qing Dynasties

The domestic market-driven economy was flourishing in the Ming Dynasty (明代), and meanwhile, foreign trade was vigorously expanding. From the third year of Yong Le (永乐) to the eighth year of Xuan De (宣德), namely, a period of about 30 years between 1405 and 1433, an official of high position in the government named Zheng He (郑和) was sent on ambassadorial voyages to the "Western Oceans" (西洋) for seven times. His Fleet reached as far as Africa and visited more than 30 countries and regions in Asia and Africa. Zheng He's voyages promoted the foreign trade and advocated the handicraft industry domestically. The technology in shipbuilding was much more advanced than that in the previous Song and Yuan dynasties (宋代, 元代). According to the *Ming Shi / Zheng He Zhuan* (《明史・郑和传》), the biggest ship in Zheng He's Fleet, 44 zhang (丈) in length and 18 zhang (丈) in width, possessed a capacity of carrying 500 to 1,000 passengers. Throughout the seven voyages, Zheng He and his accompanying officials collected and mastered large amounts of knowledge in navigation, which resulted in the well-known *Zheng He Hang Hai Tu* (《郑和航海图》, *Zheng He Navigation Maps*). Included in such maps were descriptions of technologies in measuring voyages and the depth of sea waters.

The unit used then in recording voyages was geng (更), which was about 60 li (里). The voyage of a ship of Zheng He's Fleet was about 10 geng (更) in a day and night. Affected by changes of direction of the currents, and changes of direction and speed of the wind on the ocean, the distance of 1 geng (更) could be greater or smaller than 60 li (里). The navigators in Zheng He's Fleet created an applicable method in checking the geng (更) that they were going through. A piece of wood was tied to the prow of the ship and thrown into the sea. Then a tester walked fast from the prow to the stern. The geng (更) was accurate when the speed of the piece of wood synchronized with that of the tester; the geng (更) was greater (过更) when the wood was ahead of the tester; and smaller when the wood was falling behind (不上更). Navigators in the Ming Dynasty located the locus of their ships on the seas by checking the depth of the sea waters In so doing, the unit used was tuo (庹), with 1 tuo equal to the length of the arms of an adult stretched horizontally, which was about 160 cm. The way to do it was tying a lead plummet or a heavy object and putting it in the sea waters. Some sticky stuff was applied to the plummet to get hold of what was down there at the bed of the seas. The length of the rope was then measured. And the substance stuck on the plummet was examined, whether it be sand, mud or rock, to help judge the location of the Fleet.

In the middle of the 17th century, while missionaries and merchants were arriving from the West and entering China, the people in the Qing Dynasty were enjoying an epochal prosperity in economy and unification of nationalities. Emperor Kang Xi especially paid close attention to the improvement of the celestial calendar. Noticing that the Western celestial calendar was even more scientific, he decreed to adopt the calendar *Shi Xian Li* (《时宪历》) designed by the missionary from Belgium (比利时) Ferdinand Verbiest (南怀仁) and assigned this missionary to design and produce celestial devices based on the combinations of the advanced technology from Europe and the measuring and weighing system prevailing in the Qing Dynasty. During this period of time, tremendous progresses were also made in the mapping technology. Albeit that there had been a long history with great achievements in mapping technology since the ancient times in China, precision was lacking in calculating distances, owing to the limitations in the then science and technology. In 1708, the 47th year of Kang Xi, the Royal Court decided to officially undertake a nationwide project on map designing. In so doing, advanced Western technologies and geographical knowledge were adopted. To ensure parity in measuring, the units of length were unified. It was stipulated that the standard measuring units set by Ying Zao Chi (营造尺) supervised and confirmed by the then government were the only measurement for lengths. Under this scale of units— 1 bu (步) consisted of 5 chi (尺), 1 li (里) consisted of 360 bu (步), and the distance per latitudinal

测量航速的里程单位是"更",每"更"行程约60里。为测量航速,采用简单的方法,以木片于船首投入海中,人从船首快速行至船尾,木片与人齐至,则为"更准";若木片先至,则为"过更";若人先至,则为"不上更"

The unit of mileage in measuring the speed of ships was "geng", namely, the distance was 60 li in every geng that had been elapsed. A simple method was created method in checking the accuracy of geng. A piece of wood was fixed with the rope tied to the prow of the ship and thrown into the sea. Then a tester walked fast from the prow to the stern. The geng was accurate when the speed of the piece of wood synchronized with that of the tester; the geng was greater when the wood was ahead of the tester (过更); and smaller when the wood was falling behind (不上更)

degree was 200 (里).This was the first time, and the earliest in the world, to set the length units according to the geological angular distances. It took ten years to complete the project which resulted in the National Map called *Huang Yu Quan Lan Tu* (《皇舆全览图》). Such a result symbolized the peak of the technology in the history. As was commented by a foreign scientist, it was not only the best in Asia, but was even more precise than those made in Europe at that time.

Throughout the regimes of Kang Xi(康熙), Yong Zheng (雍正) and Qian Long(乾隆), close attention was paid to agriculture. In that period of more than one hundred years, enormous projects were undertaken on irrigation, river channels, exploration in hydro-power and its applications, and they were completed with great successes. Chen Huang (陈潢) was among the top who made exceptional contributions. While redirecting the channels and changing the river beds, he utilized a method characteristic of naturally guiding the currents. To check out the actual situations of the Yellow River, he personally went on long trips along the Yellow River to collect and collate detailed first-hand data. He even went on a boat when there was a strong wind for the purpose to measure the depth of the waters. In order to forecast and control floods in time, he invented the method of measuring water, "it is 1 cubic fang (方) of water at the measurements of 1 zhang (丈) in length, width and height respectively". This was applied to calculating the volume of water flowing through in certain unit of time. He said confidently that it is possible to check out the amount of water flowing through in a day. The achievements Chen Huang made in hydrology was closely associated with his method of water measuring.

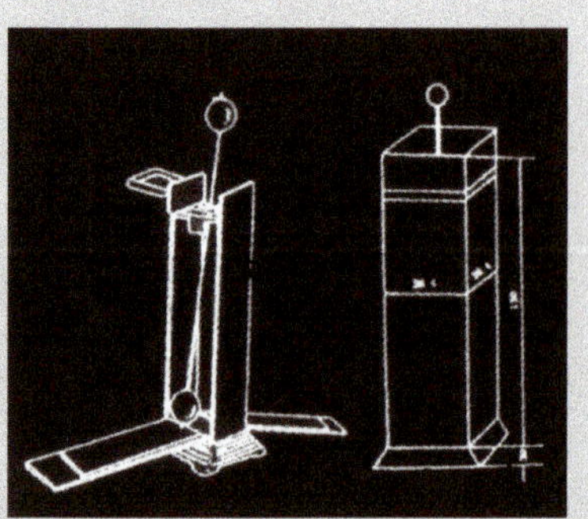

六合验时器
A 6-he time reckoning instrument

清代 康熙年间制造的六合验时器（摆动周期为1秒），用它测定河道闸口处水的流速，再乘以闸口面积，以测算闸口水的流量
A 6-he time reckoning instrument invented during the years of Kang Xi in the Qing Dynasty (flickering cycle: 1 second), used to determine the flow rate per second at the water gate of the river. The flux was reached by calculating the flow rate times the area of the water gate

清代　陈潢在治理水患中，为查明水情"驾轻舟，测水之深浅"

Chen Huang bringing floods under control in the Qing Dynasty.In order to check out the water regimen, he "was propelling a boat to determine the depth of the waters".

在四川涪陵长江河道中有一块巨大的礁石，上面记刻从唐代至中华民国时期长达1200多年的长江枯水位。这是中国自古对水位测量、记录的见证文物

There was a huge rock in the streamway of the Yangtze River in Fu Ling, Sichuan Province, on which marks were carved to indicate the low water levels throughout the 1,200 years from the Tang Dynasty to the Republic of China. This is a historic relic of evidence in measuring and recording water levels since ancient times in China

清代　记录水位的铁牛

An iron bull recording water levels in the Qing Dynasty

THE HISTORY  OF ANCIENT CHINESE

12

THE HISTORY OF ANCIENT CHINESE MEASURES AND WEIGHTS

中国古代计量史

EASURES AND WEIGHTS

十二、近代对度量衡的改革及米制的推行

Chapter Twelve

Innovations on the Measuring and Weighing System and the Implementation of the Metric System

清代末年政治腐败，当权者对内无能治理国家，对外无力抵御帝国主义的入侵。1840 年鸦片战争之后，中国逐步沦为半殖民地半封建社会。由于外国经济的侵略，各国度量衡也纷纷传入，对中国度量衡产生了直接的影响，使它带有很深的半殖民地性质。清政府既无力抵制各国制度在中国的使用，也无力统一国内的度量衡，因此造成了制度、器具、量值等各方面的极度混乱。

Towards the last days of the Qing Dynasty (清代), the politics became notorious owing to the prevailing corruption in the then government. Those in power were incapable for governing the state, and what was even worse was that they did not know where to turn in front of imperial invaders from outside. With the Opium War (鸦片战争) breaking out in 1840, China gradually declined to a semi-colonial and semi-feudal society.

Landing on our land together with the invasions were the measuring and weighing systems of foreign countries, which began to affect unfavorably the long existing domestic measuring and weighing system. The consequences were obvious as the domestic system appeared to carry a semi-colonial characteristic. The Qing government was neither having the ability to resist the influx of those foreign systems and their applications in the domestic daily life, nor possessing the power to unify the measuring and weighing systems in China. Hence there appeared again tremendous chaos in regularity stipulation, apparatus making and value setting.

清末 英国驻广州通商口岸领事馆的英制砝码
English weights kept in the British Consulate in the Guangzhou Treaty Port, at the end of the Qing Dynasty

清代 铜砣 錾刻“计司马秤五斤，合英权七磅正”。重 2650 克
A bronze sliding weight, 2,650 g in weight, in the Qing Dynasty, engraved with “5 Jin for the Sima Steelyard, Exactly Equal to 7 English Pounds”

清代 铜砝码 一面刻“格拉姆(姆)”。“格拉姆”是米制重量单位 Gram 的中文译名。重 37.3 克，正合库平一两
A bronze weight, 37 g in weight, just equal to 1 Ku Ping liang, in the Qing Dynasty, carved on one side with “Ge La Mu” (Gram), a transliteration in Chinese of a metric unit of mass in the metric system

1. 海关主权的丧失与海关度量衡的产生

道光(1821 ～ 1850)后期，中外通商日益频繁。为了便于稽查和征收进出口货税，清政府设立了通商海关。根据道光二十七年(1847)中国、瑞典、挪威《贸易章程》第十二款规定，瑞典、挪威等各国口岸领事馆所有度量衡，均以粤海关定式为标准。由“中国海关发给丈尺、秤码各一副，以备丈量长短权衡轻重之用”。广州、厦门、福州、宁波、上海等五个通商口岸，一律依照粤海关所颁之式盖戳镌字，以保证量值统一。并要求各海关监督所在口岸送交领事馆照此统一。然而各国政府并不满足于清政府开放海禁，为了在中国攫取更多的资产，各国都想控制中国的海关大权。咸丰八年(1858)《天津条约》订立之后，各约所附通商章程皆规定请外国人帮助办理税务，从此海关大权旁落。咸丰九年(1859)英国人李泰国(Horatio Nelson Lay)攫夺了广东海关后，中国的海关一直被帝国主义把持。海关衙门既不在中国行政系统之内，海关上的度量衡以及币制也不在中国法律规定之内。各国海关衙门借口中国度量衡庞杂纷乱，另设专款规定相互折合的办法。自此在中国度量衡史上出现了丧权辱国的海关度量衡，即所谓的海关尺和关平秤。

清末荷兰东印度公司用英制砝码称从中国出口的茶叶

English weights used by the Eastern Indian Inc. of the Netherlands to weigh the tea exported from China at the end of the Qing Dynasty

清末广州通商口岸
The Guangzhou Treaty Port towards the end of the Qing Dynasty

《权度法》公布后，在农商部下设立权度局和权度检定所、权度制造所。此铁砣铸阳文"权度制造所造"
After *the Law on Weights and Measures* was promulgated, the Bureau of Weights and Measures, the Institute of Certifying Weights and Measures, and the Institute of Producing Instruments of Weights and Measures were established under the Ministry of Agriculture and Commerce. This is an iron sliding weight with characters cut in relief as "the Institute of Producing Instruments of Weights and Measures"

1. The Relinquishment of Sovereignty in Custom and the Emergence of Measuring and Weighing System Applied in Custom Inspection

During the last period of Dao Guang (道光, A. D. 1821-1850), the frequency in foreign trade was increasing rapidly. In order to inspect and impose tariff on imported and exported goods, the Qing government set up the Custom Inspection and regulate the measuring and weighing system. According to the regulations stated in Item 12 in *the Foreign Trade Charter* (《贸易章程》) granted in 1847, the 27th year of Dao Guang (道光), the standard units for measuring and weighing imported and exported goods determined by the Canton Custom (粤海关) must be observed by the Swedish (瑞典) and Norwegian High Commissions and others alike. The Custom of China distributed rulers (丈尺) and weights (秤码), one set of each, to the customs in Guangzhou, Xiamen, Fuzhou, Ningbo and Shanghai, for measuring and weighing the imported and exported goods. The inspection was ordered to carry out with standardized formats of stamp and seal as implemented in the Canton Custom, to ensure a unified system in values. Those Customs were requested to supervise the reports sent to the pertinent high commissions, for the sake of parity. Unexpectedly, the governments of the related countries were not satisfied with the "open-door policy", and, for the purpose of seizing more opulence from China, they intended to control the Custom of China. In 1858, the 8th year of Xian Feng (咸丰), *the Tianjin Treaty* (《天津条约》) was signed. According to the items stated in this Treaty, and the regulations attached to it, foreigners were invited to help with the tariff imposed on the imported and exported goods, thus the authority

1912 年中华民国成立，袁世凯任第二届临时大总统。1915 年 12 月他下令恢复帝制，改国号"中华帝国"，次年改为洪宪元年。为此而发行纪念币和烧制瓷器。这件"中华帝国乙卯年(1915)造"瓷砣，正反映了这一史实
The Republic of China was founded in 1912, and Yuan Shikai took the power as the Temporary President of the 2nd presidential term. In December, 1915 he ordered to resume the monarchy, and to transform the title of the regime as "Zhong Hua Di Guo" (中华帝国), and further changed the title of the regime as "The 1st year of Hong Xian" in the following year. To consolidate his rule, commemorative bank notes were issued and new chinaware was baked. This china sliding weight stamped with "Made in the Year of Yi Mao of Zhong Hua Di Guo (1915)" reflected the above historical facts

of tax levying was in the hands of foreigners. In the 9th year of Xian Feng, an Englishman named Horatio Nelson Lay seized the Canton Custom. The Custom of China was controlled by foreigners ever since then. The divisions of the customs were not under the Chinese administrative systems, and the measuring and weighing systems, together with the currency exchanges, were not stipulated in accordance with the Chinese laws and regulations. The worst part was that the foreigners working in the customs took the excuse that the Chinese measuring and weighing systems were too complex and chaotic, and there were no standards to observe, therefore, they had the right to establish regulations on their own, including the exchange rates. From then on, for the first time in the Chinese history of measuring and weighing system, there appeared the ashamed "measuring and weighing system for customs" (海关度量衡), known as the so-called Custom Ruler (海关尺) and Custom Cheng (关平秤).

根据《度量衡法》规定，工商部设立全国度量衡局，地方上设立度量衡检定和制造所。此为各级地方政府制造的秤砣，都铸有阳文标志

In accordance with the regulations stipulated in *The Law on Measuring and Weighing Systems*, the National Bureau of Measuring and Weighing Systems was set up under the Ministry of Industry and Commerce, and the Institute of Producing and Certifying Instruments of Weights and Measures was to be established in each of the local authorities. Hence all the steelyard weights produced at all levels of the local authorities carried with identification codes cut in relief

2. 营造库平制与米制并用的时期

随着清政府被推翻，度量衡的改革也进入一个新时期。从国际形势来看，自 1791 年法国政府颁布米制的标准后，世界上采用米制的国家日益增多。1875 年 3 月 1 日，法国政府召集了"米制外交会议"，20 多个国家派出了政府代表与科学家出席会议，正式签署了《米制公约》。国内形势是，鸦片战争后，中国度量衡无统一的准则，制度混乱。1912 年中华民国成立，正是彻底改革的极好时机。工商部经反复讨论，均认为适应世界潮流，直接采用米制，既可统一全国混乱的制度，又可消除对外贸易的障碍，并拟在 10 年内将米制推行全国。后因考虑米尺过长，公斤(千克)过重，完全废除营造库平制，直接采用米制，恐与民情习俗不合而不易被接受。故采用美、英等国将本国度量衡与米制兼用的办法，确定营造库平制与米制并用。1915 年，《权度法》颁布，明确了米制与营造库平制的比例和折算关系，并设立了权度制造所和权度检定所，力争推行新法。然而由于当时政局不稳、号令不行、计划不周、经费不足等原因，新制实际上并未能在全国推行。

中华民国四年(1915),政府公布《权度法》,采用甲制(营造库平制)、乙制(万国权度公制)并用。并商请国际计量局制造公斤(千克)原器(重999.91克)和库平两原器(重37.3克)

In the 4th year of the Republic of China (1915), the then Government promulgated *Law on Weights and Measures,* to adopt the Jia System (Ying Zao Ku Ping Zhi) and Yi System (International Metric System). As requested, the then International Bureau of Measures and Weights produced the Standard Instrument of Kilogram (999.91 g) and the Back-up Ku Ping Liang (37.3 g)

A.公斤(千克)原器,重999.91克
The Standard Instrument of Kilogram, 999.91g in weight

B.库平两原器,重37.3克
The Back-up Ku Ping Liang, 37.3g in weight

2. The Period When Ying Zao Ku Ping Zhi (营造库平制) Co-existed with the Metric System

After the Qing Dynasty was toppled down, China witnessed a new era with innovations on the measuring and weighing system. From the international perspective, ever since the standards of metric system (米制) was granted by the then French Government in 1791, more and more countries adopted such a system. On March 1, 1875, the French Government organized an international conference on the metric system, and representatives and scientists were sent from more than twenty countries to attend it. It was at this Conference that *the Metric System Treaty* (米制公约) was signed officially. Domestically, after the Opium War, the situations were very chaotic because there were no standards to follow in terms of the measuring and weighing systems. At the beginning of the Republic of China (中华民国, 1912), an ideal opportunity turned up to innovate the systems thoroughly. After many rounds of discussion in the Department of Industry and Commerce (工商部), an irresistible opinion was that the metric system ought to be adopted, to follow the international trends. The dual merits lay in that it would not only re-unify the systems from the chaos, but also would remove the hurdles in foreign trade. It was also planned to implement the metric system all over China within ten years. Nevertheless, in considering that 1 meter was too long in length, and 1 kilogram was too heavy in weight, people would feel it hard to accept the new system when they saw Ying Zao Ku Ping Zhi (库平制) all exterminated from their daily life. Therefore, it was proposed that Ying Zao Ku Ping Zhi co-exist with the metric system, just like the way applied and applicable both in the United Kingdom and the United States, to continue their traditional system while adopting the metric system at the same time. In 1915, *the Law on Weights and Measures* (《权度法》) was granted. It articulated the exchange relations between Ying Zao Chi Ku Ping Zhi (营造尺库平制) and the Metric System. Meanwhile, the Institute of Producing Instruments of Weights and Measures (权度制造所) and the Institute of Certifying Weights and Measures (权度检定所) were in position for a forceful implementation of them. However, owing to the turmoil in the political arena, together with such factors as difficult implementation, inconsistent planning and lack of funds, the new systems were not implemented officially.

中华民国时期 铜砝码 1929年,政府废除《权度法》,颁布《度量衡法》。其中第二条规定,采用"公制"为标准制,并暂设"市用制"为辅制。这两件"二十公斤(千克)""十五公斤(千克)"铜砝码原保存在全国度量衡局,当属标准器

Bronze weights in the years of the Republic of China.In 1929, the then Government abrogated *the Law on Weights and Measures*, and promulgated *the Law on Measuring and Weighing Systems*, and it was regulated in Article 2 that the "Metric System" was adopted as the standard, and that the "Domestic System" was used temporarily as the back-up system.The two bronze weights, "20 kg" and "15 kg" respectively, originally retained in the National Bureau of Measuring and Weighing Systems, were the standard instruments at that time

3. 市用制确立与米制推行

1927 年南京政府成立。由于度量衡关系到国家政治及人民福利，再次提出统一度量衡议案。随即组织度量衡标准委员会进行度量衡标准的研究，先后提出了十几个方案，最后决定采用徐善祥、吴承洛的提案：即拟采用米制，在过渡时期宜采用与米制的比率为 1 米等于 3 市尺、1 公升等于 1 市升、1 公斤（千克）等于 2 市斤的市用制。由于市用制既概略地沿用了营造库平制的量值，又与米制有简单的比率。经会议反复讨论后，一致同意此提案，于 1928 年制定《中华民国权度标准方案》，予以公布。为了保证方案的实施，由中央各部与全国商会联合组织度量衡推行委员会，制定了全国度量衡统一程序，公布《度量衡法》。根据《度量衡法》将全国各区域按交通、经济发展的不同情况分区域、分期逐步推行米制和市制。然而，由于国民经济衰退，工业、科学技术以及教育事业凋零，米制未能在全国推行，只是与人民生活有密切关系的市用制在各省市商业行会中逐步通行起来。

中华民国时期度量衡统一活动，有效地遏制了清末民初以来度量衡的混乱状况，为我国度量衡制度与国际接轨，为以后推行米制奠定了基础。

中华民国时期 铁砝码 铸铭“山西省公署一斤”为省级市用制标准器
An iron weight made in the Republic of China cast with “1 Jin, Prefecture of Shanxi Province”, which was the standard instrument of the Domestic System at the provincial level

中华民国时期 “1 公斤（千克）”“1 市斤”铁砝码
Iron weights of “1 kg” and “1 jin” in the Republic of China

1947 年中华民国政府成立经济部取代实业部，将度量衡局与标准委员会合并，成立中央标准局。此铁秤砣铸铭“经济总署度量衡制造工厂”“秤量三百市斤”
In 1947, the then Government set up the Ministry of Economy to replace the Ministry of Industry, and combined the Bureau of Measuring and Weighing Systems and the Commission for Standardization, to establish the Central Bureau of Standardization. This iron steelyard weight was engraved with “The Factory Making Measuring and Weighing Instruments under the General Administration of Economy” and “For Steelyards to Weigh up to 300 Jin”

中华民国时期 铁砣 铸阳文“普通”“公权”。重 7150 克
An iron steelyard weight, 7,150 g in weight, in the Republic of China, with “Pu Tong” and “Gong Quan” cut in relief

中华民国时期 铁砣 铸阳文"公粮""市秤"
An iron steelyard weight in the Republic of China, with "Gong Liang" and "Shi Cheng" cut in relief

中华民国时期 铁砣 铸阳文"甲子年(1924)立""经布平"。重 9800 克
An iron steelyard weight, 9,800 g in weight, in the Republic of China, with "Produced in the Year of Jia Zi (1924)" and "Jing Bu Ping" cut in relief

3. The Establishment of the Domestic System and the Implementation of the Metric System

In 1927, the Nanjing Government (南京政府) was founded. An act was proposed to unify the measuring and weighing systems again, because they were closely related to the interests of the State and to people's daily life. The Committee of Standards for Measures and Weights (度量衡标准委员会) was set up

中华民国时期 铁砣 铸阳文"较准官秤""呼兰糖厂"。重 6900 克
An iron steelyard weight, 6,900 g in weight, in the Republic of China, with "Officially Standard Steelyard" and "Hu Lan Tang Factory" cut in relief

to undertake researches on the systems. There were more than a dozen proposals put forward for discussion, and the Proposal by Xu Shanxiang (徐善祥) and Wu Chengluo (吴承洛) was adopted and decided on. In this Proposal, it was advocated to adopt the Metric System (米制), and in the meantime, since a transitional period had to be experienced, a system of exchange rates was set, as follows. 3 shi chi (市尺)=1 m (meter，米),1 shi sheng (市升)= 1L [liter，公升 (升)], 2 shi jin (市斤)=1 kg [kilogram, 公斤(千克)]. In light of the merits of the Proposal that the new Domestic System (市制) carried on the values of the traditional Ying Zao Ku Ping Zhi (营造库平制) and that it provided a simple rates in exchange with the Metric System, the Proposal was passed in consensus. In 1928, *the Standards for Weights and Measures of the Republic of China* (《中华民国权度标准方案》) was granted. To make sure that it was properly implemented, all the departments of the then Central Government and the National Association of Merchandise formed up the Measures and Weights Implementation Committee (度量衡推行委员会) that specified the stages and processes for the nationwide re-unification of the measuring and weighing system. Following this was the granting of *the Law of Measuring and Weighing Systems* (《度量衡法》). According to the Law, the country was divided into various regions based on the economic development, transportation and so on, to adopt the new Domestic System (市制) together with the Metric System step by step. Notwithstanding that the processes were well planned, the Metric System was not carried out because of the declining economy, troublesome situations in industry, education and in science and technology. The new Domestic System, understandably, was implemented gradually among merchandise in provinces across the country, since the new system was closely related to the daily life of common people.The efforts exerted on re-unifying the measuring and weighing systems during the period of the Republic of China effectively turned around the confusing situations from the last days of the Qing Dynasty, and laid a solid foundation for the development of our measuring and weighing systems, at a later stage in the new era, to be connected with the international trend, that is, to carry out the Metric System.

中华民国时期　工商部全国度量衡局制造所制的一组标准量器

A set of standard instruments of capacity produced by the National Bureau of Measuring and Weighing systems under the Ministry of Industry and Commerce in the Republic of China

中华民国时期　一升铜量 錾刻“一公升＝一升”

A 1-sheng bronze instrument for capacity in the Republic of China, engraved with “1 Liter = 1 Sheng”

中华民国时期　五升铜量

A 5-sheng bronze instrument for capacity in the Republic of China

中华民国时期　一斗铜量

A 1-dou bronze instrument for capacity in the Republic of China

中华民国时期　一升铬量 錾刻“工商部全国度量衡制造所”“一升”

A 1- sheng chromium instrument for capacity in the Republic of China, engraved with “The National Bureau of Measuring and Weighing Systems under the Ministry of Industry and Commerce” and “1 Sheng”

中华民国时期　木斗

A wooden dou in the Republic of China

中华民国时期　木升

A wooden sheng in the Republic of China

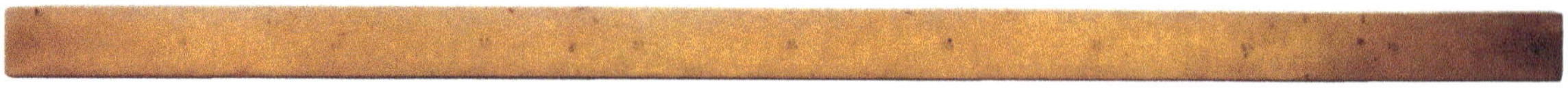

中华民国时期 铜尺 錾刻“农商部权度制造所制”“同 269”。长 32 厘米
A bronze ruler, 32 cm in length, in the Republic of China, engraved with “Produced by the Institute of Producing Instruments of Weights and Measures under the Ministry of Agriculture and Commerce” and “Tong 269”

中华民国时期 铜尺 錾刻“经济部全国度量衡制造所”。长 33.3 厘米
A bronze ruler, 33.3 cm in length, in the Republic of China, engraved with “Produced by the National Institute of Measuring and Weighing Systems under the Ministry of Economy”

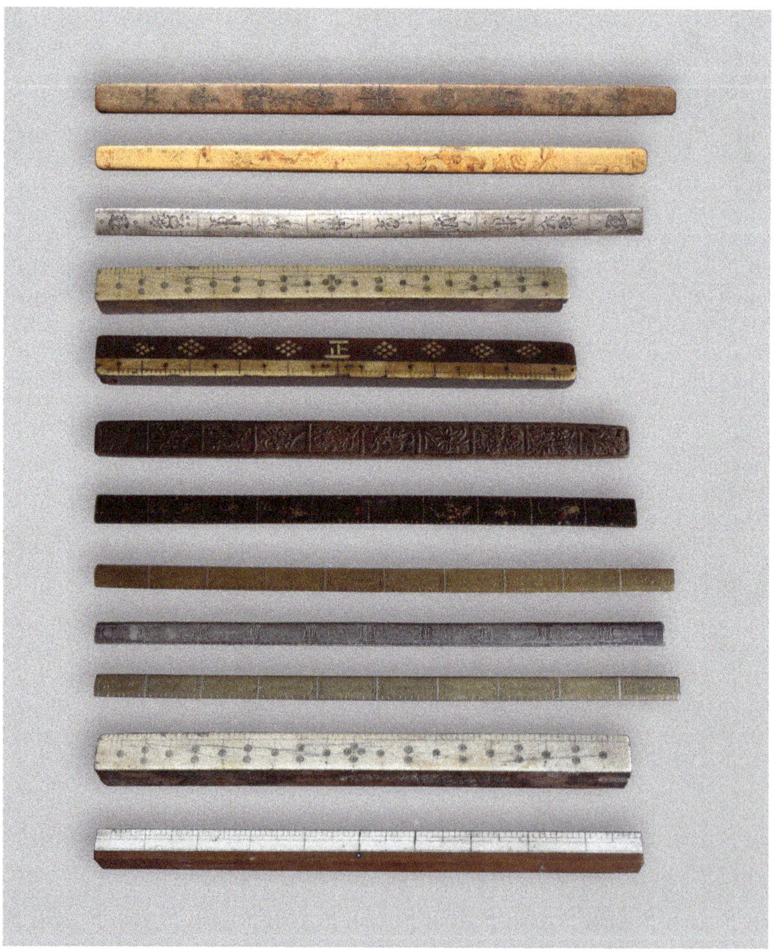

中华民国时期 一组民间使用长短不符标准的尺
A series of rulers with irregular lengths, non-standard but commonly used in the Republic of China

中国历代度量衡量值表

A Chart of Sizes of Chinese Measures and Weights of All the Past Dynasties

时代	年代（公元）	单位量值		
		一尺合厘米数	一升合毫升数	一斤合克数
商	前 1600~前 1046	16		
战国	前 475~前 221			
（齐）			205	370/镒
（邹）			200	
（楚）			226	250
（魏）			225	306/镒
（赵）			175	251
（韩）			168	
（东周）		23.1	200	1213/寽
（燕）			1766/觳	251
（中山）			180	9788/石
（秦）		23.1	200	253
秦	前 221~前 206	23.1	200	253
西汉	前 206~8	23.1	200	250
新	9~23	23.1	200	245
东汉	25~220	23.1	200	220
三国	220~280	24.2	200	220
晋	265~420	24.2	200	220
南北朝	420~589			
（南朝）		24.7	200	220
（北朝）		25.6~30	300（前期）	330（前期）
			600（后期）	660（后期）
隋	581~618	29.5	600	660
唐	618~907	30.3	600	667
宋	960~1279	31.4	702	640
元	1206~1368	35	1003	640
明	1368~1644	32	1035	596.8
清	1616~1911	32	1035	596.8
中华民国	1912~1949	33.3	1000	500

Postscript

THE HISTORY OF ANCIENT CHINESE MEASURES AND WEIGHTS

后 记

1975 年，在国家计量总局领导下，成立了“度量衡史料小组”，打算编辑一本《中国古代度量衡图集》。这项工作得到了国家文物局的支持。由国家计量总局、中国历史博物馆、故宫博物院等部门的有关人员组成的编辑小组分赴全国各地博物馆和文物管理部门，作了一次古代度量衡器全面的调查、搜集、整理和研究工作。1977 年该书初稿完成后，在国家计量总局的主持下，召开了该书的审稿会，邀请全国各地有关学者参加，得到专家们的全力支持。当年与会的著名学者有：夏鼐、唐兰、商承祚、王振铎、张政烺、罗福颐、朱德熙、杨宽、史树青、李学勤、马承源、裘锡圭、俞伟超、马雍、杨伯达、王世民等。在长达 7 天的会议中，专家们对书稿作了仔细的推敲、反复的审定，提出了具体、重要的修改意见。会后，史树青先生更是不惜时间和精力，主持编辑组对每一件器物文字说明的讨论和修改。最后，全部文字稿由当时任北京大学副校长的朱德熙先生亲自审订。如果没有这些专家的指导和帮助，就不会有 1980 年由文物出版社出版的《中国古代度量衡图集》，就不会有此后我们对中国计量史研究工作的继续。30 多年过去了，当年与会的大多数学者已先后仙逝，为了铭记前辈的扶持，也为了不忘记这段历史，在此不惜篇幅地记录下来。

《中国古代计量史》在编辑过程中，同样得到各方面人士的帮助和支持。在搜集器物彩色图片时，前中国历史博物馆藏品室主任马秀银女士多次提供她珍藏的资料；中国科学院科学技术史所资料室李映新先生等，更为我选录素材给予大力支持和帮助。书中许多未曾发表的器具，多由山东省计量科学研究院、北京“古权斋”物主郑海生先生提供，照片多由李云飞、杨路滨先生拍摄。书名由书画家陈青子女士题写。书中插图多由吴海先生创作。文字则由邱隆先生审阅。在此一并致以衷心的感谢。

更让我感到荣幸的是，原中国科学院资深院士王大珩先生欣然为本书作序。王老对计量事业的关怀，对后辈扶持的崇高风尚，让我深受教益。

www.ingramcontent.com/pod-product-compliance
Lightning Source LLC
Chambersburg PA
CBHW060937200326
41128CB00013B/108
9781921816031